THERMOMETRIC AND ENTHALPIMETRIC TITRIMETRY

VAN NOSTRAND REINHOLD
SERIES IN ANALYTICAL CHEMISTRY

Edited by

DR. R. A. CHALMERS

Department of Chemistry
University of Aberdeen

This series is designed as a coverage of reliable analytical information of value to chemists in research, industry and teaching. Each volume is carefully selected and planned as a modern treatment of a topic of importance to analytical chemists today. New volumes will be added from time to time.

Chemistry of Complex Equilibria	M. T. BECK
Solvent Extraction of Metals	A. K. DE, S. M. KHOPKAR and R. A. CHALMERS
Modern Methods in Organic Microanalysis	JEAN P. DIXON
Errors, Measurement and Results in Chemical Analysis	K. ECKSCHLAGER
Laboratory Handbook of Chromatographic Methods	O. MIKEŠ
Flame Photometry Theory	E. PUNGOR
Solution Equilibria in Analytical Chemistry	L. ŠŮCHA and S. KORLÝ
Thermal Analysis	A. BLAŽEK
Inorganic Chromatographic Analysis	J. MICHAL
Gas-Liquid Chromatography	D. W. GRANT
Atomic Fluorescence Spectroscopy	V. SYCHRA

Thermometric and Enthalpimetric Titrimetry

G. A. VAUGHAN

*Chief Analyst, Lloyds' Pharmaceuticals
Clerk Green, Batley, Yorkshire*

VAN NOSTRAND REINHOLD COMPANY
LONDON
NEW YORK CINCINNATI TORONTO MELBOURNE

VAN NOSTRAND REINHOLD COMPANY LTD.
25–28 Buckingham Gate, London, SW1E 6LQ

INTERNATIONAL OFFICES
New York   Cincinnati   Toronto   Melbourne

Library of Congress Catalog Card No. 79–186764
ISBN 0 442 78385 X

First published 1973

Printed in Great Britain by Butler & Tanner Ltd, Frome and London

Contents

Preface

The fact that temperature changes take place when substances are mixed together is one of the first observations that any aspiring young chemist will have made, but few chemists know of the use that can be made of this temperature change, both in analysis and in thermodynamics.

One of the main causes of temperature change is the heat evolved or absorbed in a reaction and because this is proportional to the enthalpy of the reaction and to the extent that the reaction has progressed, it follows that temperature changes can be used for analytical purposes. This was recognized in 1910 when Howard analysed fuming sulphuric acid by adding water to it in the form of dilute sulphuric acid and measuring the temperature rise, and later in 1913 when Bell and Cowell prepared neutral solutions of ammonium citrate by adding ammonia solution to citric acid solution until the temperature of the mixture ceased to rise. These early examples illustrate the two ways in which the heat of reaction can be utilized; in the first, an enthalpimetric titration, the temperature change is itself a measure of the reaction; in the second, a thermometric titration, the temperature change is used to determine the end-point of the reaction.

This dual nature of the technique has given rise to confusion in the nomenclature to be used: '. . . "thermometric titration" should not be used. . .' (Hallet, 1952); '. . . "enthalpy titration" should be used. . .' (Jordan and Alleman, 1957); '. . . "thermometric titrations" should be adhered to in the future. . .' (Hume and Jordan, 1958). In this book the principle of Mellon (1955) viz. 'that analytical nomenclature should consistently identify the procedures by the characteristic property involved' has been adhered to so that the word 'thermometric' or 'enthalpimetric' is used whichever is appropriate. The titration graph has been described as such and not as a thermogram because the latter term is used for the results obtained by the medical technique of thermography (see *Nomenclature in Thermal Analysis*, Mackenzie, 1969).

In very many cases the enthalpy of the reaction is known and the heat and thus the temperature change in any titration can be calculated, but where these values are not available, T values in degC/mole have been calculated from the titration graphs given in the papers, and these approximate values are included to help those wishing to use a particular method. The calorie heat unit and not the S.I. unit, the Joule, has been used throughout the text because it has been used in the majority of papers so far published. The conversion 1 J = 0·239 cal or 1 cal = 4·187 J can be used for comparing results from future papers in which S.I. units may be used. Because the subject is concerned with both change in temperature and degree of temperature, to avoid the confusion that could arise, the British Standard 1991 of degC and degK has been used throughout the text to denote temperature interval whilst °C and °K have their usual meanings of temperature value.

The history of the technique falls into two phases; the first, in which simple apparatus and laborious procedures were used, is described in the text as 'the early titration method' to distinguish it from the second, modern phase, in which thermistors or other sensitive temperature detectors are used in conjunction with motor-driven syringe burettes or peristaltic pumps, a combination that enables the temperature changes to be recorded. It is this reliable instrumentation, first described by Linde, Rogers and Hume in 1953, that has caused the resurgence of interest in the method and the widespread increase in its use.

It is hoped that this book will help chemists, particularly those engaged in analysis, to appreciate the scope of this remarkable technique. The contents have been arranged so that it should be relatively easy to compare the methods of analysis or interactions of a particular metal or functional group.

I wish to acknowledge the help given by the library staff of the Coal Tar Research Association and of Lloyds' Pharmaceuticals and the invaluable assistance given by my wife in the preparation of the manuscript. I am also most grateful to the copyright holders for the permission so readily and generously granted to reprint tabular material and figures.

Chapter 1

Theoretical Aspects

1.1 Introduction

Calorimetric procedures can be conveniently divided into two classes, which differ in the manner in which the reactants are introduced into the adiabatic system or calorimeter. In conventional calorimetry the total quantity of each reactant is placed in the calorimeter before the reaction is initiated. The rate of reaction from then on is controlled by the kinetics of the reaction and by the diffusion of the reactants or by the method of mixing.

In thermometric or enthalpimetric titrimetry one reactant is placed in, or added continuously to, the calorimeter and the other is added to it or generated within it, either continuously or incrementally, the degree of reaction being controlled by the rate of the addition or generation. These titrations can therefore be defined as those in which the temperature of the fluid being titrated is measured or recorded during the titration. The measured temperature can be utilized in two distinct ways and this distinguishes between thermometric and enthalpimetric titrations, which can be defined as follows:

Thermometric titration   A titration in which the temperature of the fluid titrated is used to determine the end-point of the titration.

Note that this definition does not exclude post-titration thermometric indicator systems (see Sections 1.4.2 and 1.8.4).

Enthalpimetric titration   A titration in which the change in temperature caused by the reaction is determined.

It follows from these definitions that heat changes are an integral part of both thermometric and enthalpimetric titrimetry.

1.2 Heat Changes

Heat changes that take place in chemical systems can be conveniently considered in two groups, (*i*) changes of state and (*ii*) interaction.

In a single-component system the heat changes are:

(*a*)   heat of vaporization or sublimation
(*b*)   heat of fusion
(*c*)   heat of transition

(*d*) heat of formation

and (*e*) heat of temperature change (specific heat).

Additionally, in multiple-component systems the following heat changes apply:

(*f*) heat of solution
(*g*) heat of dilution or mixing
and (*h*) heat of reaction.

Titrimetry is a multiple-component operation and thus thermometric and enthalpimetric titrimetry is concerned with these three types of heat change and in particular with the heat of reaction, though other heat changes, e.g. heat of vaporization, may be present during the titration.

1.3 Heat of Reaction

From thermodynamic studies it is known that the heat change in any chemical reaction at constant pressure in a particular system is composed of three parts, which are related as shown in Eq. 1

$$\Delta H = \Delta G + T \Delta S \tag{1}$$

where, in the reaction, ΔH is the change in enthalpy, ΔG is the change in free energy, ΔS is the change in entropy and T is the temperature in Kelvin.

If the reaction has the form shown in Eq. 2,

$$aA + bB \rightleftharpoons pP \tag{2}$$

and the reaction, when complete, produces a molar heat of reaction of ΔH which is shown as a measurable temperature change, ΔT, in the system, then the total heat evolved in the reaction, Q, is related to ΔH and ΔT as shown by Eqs. 3 and 4

$$Q = -n_P \Delta H \tag{3}$$
$$Q = C_s \Delta T \tag{4}$$

where n_P is the number of moles of product P formed and C_s is the heat capacity of the system.

Combining Eqs. 3 and 4 gives Eq. 5 that relates ΔT and ΔH

$$\Delta T = -\Delta H \, n_P / C_s \tag{5}$$

It follows from Eq. 1 that if the value of the entropy term, $T \Delta S$, does not negate the free-energy term, ΔG, then ideally from Eq. 5, ΔT will be a measure not only of the total molar heat of reaction but also of the number of moles that have reacted.

Table 1.1 gives the temperature change, ΔT, that can be expected for different combinations of the two variables (*i*) the molar concentrations of the product, n_PM, and (*ii*) the molar heats of the reaction, ΔH. In calculating these theoretical temperature changes from Eqs. 3 and 4 it is assumed that the density and specific heat of the medium have a value of unity.

Table 1.1 Temperature change in a titration medium for different values of molar concentrations and molar heats of reaction

(From Sajó and Sipos (1967b))

	Heat of reaction, ΔH, kcal/mole			Temperature change, ΔT
	± 10	± 1	± 0.1	$\mp$degC
Molarity	1	—	—	10
of	0.1	1	—	1
the	0.01	0.1	1	0.1
product	0.001	0.01	0.1	0.01
n_PM	0.0001	0.001	0.01	0.001
	0.00001	0.0001	0.001	0.0001

The three terms of Eq. 1 are practically independent of each other so it is possible for an investigation of a reaction to be unsuccessful because it is based on free-energy measurements, whereas if there is a favourable entropy change then measurements based on total energy should succeed. This is exemplified in the neutralization titration of boric acid in which the indicator titration method, based on a free-energy measurement, fails, whereas the thermometric titration succeeds (see Neutralization Reactions, Section 3.48).

1.3.1 *Basis of thermometric and enthalpimetric titrimetry*

If the general reaction given in Eq. 2 is considered, where the two components A and B react to form a product P and this produces a temperature change of ΔT in the system, then it follows that if B is added continuously or incrementally to A there will be a continuous or stepwise positive or negative change in temperature of the system until B is in excess, when the temperature will cease to change. This cessation marks the end-point of the reaction, i.e., the addition is a thermometric titration in which the absolute amount of A is measured. If B is added all at once so that it is in excess, then from Eq. 5 the temperature change, ΔT, of the system will be a measure of the number of moles that have reacted, i.e., the addition is an enthalpimetric titration in which the concentration of A is measured.

The plot of the temperature of the system against the time in which a volume of titrant is added is called the titration curve or enthalpigram and the idealized forms of these are shown in Fig. 1.1.

It can be seen in Fig. 1.1 that in a thermometric titration, whether the titrant is added continuously or incrementally, the addition must be regular with respect to the time over the whole of the titration period, and in this way a sharp inflection is obtained in the titration curve between the reaction period (b) and the excess titrant period (c) that marks the end-point of the titration. In an enthalpimetric titration, the titrant must be added as rapidly as is possible so as to obtain a true heat 'pulse'.

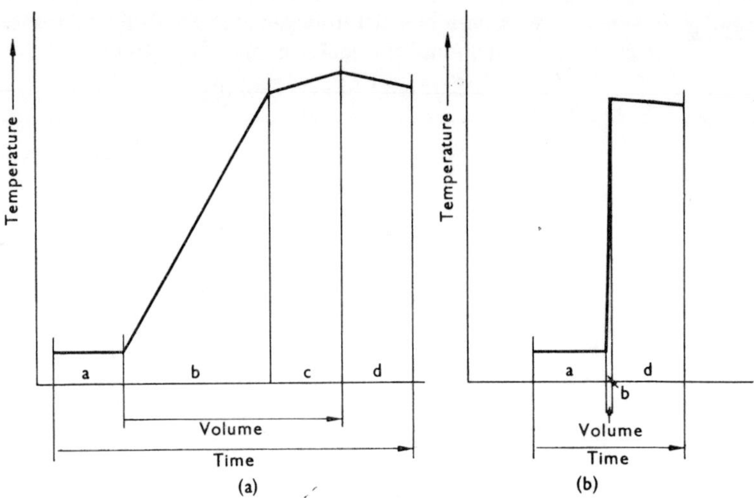

FIG. 1.1 Idealized forms of thermometric (a) and enthalpimetric (b) titration curves for an exothermic reaction.

(a) base period—no titrant added;
(b) reaction period—titrant added;
(c) excess titrant period—titrant added;
(d) post-titration period—no titrant added.

1.3.2 *Significance of the equilibrium constant*

The free-energy term, ΔG, of Eq. 1 (referred to the standard state, i.e. $\Delta G°$) can be related to other parameters of the reaction:

$$\Delta G = -RT \ln K \tag{6}$$

where R is the gas constant, T is the temperature in Kelvin and K is the equilibrium constant of the reaction.

In the discussion above no account has been taken of the completeness or otherwise of the reaction between A and B. This completeness is expressed by the equilibrium constant K of Eq. 6 and is related to the concentration of the reactants according to Eq. 7

$$K = \frac{[P]}{[A][B]} \times \frac{\nu_P}{\nu_A \nu_B} \tag{7}$$

where ν is the appropriate activity coefficient.

If the reaction of B with A is not complete near the equivalence point, less heat will be evolved per unit of B added and the slope of a thermometric titration curve will change and become more curved about the equivalence point however high the heat of the reaction, the curvature being greater the smaller the value of the equilibrium constant K. However, the general property of equilibrium curves is such that tangential extrapolation from both sides of the curved region will give the non-equilibrium curve, and the intersection is the end-point of the reaction in the case of thermometric titration curves.

Similar considerations apply to an enthalpimetric titration, but as the object of this titration is to produce as sharp a temperature rise as is possible, it is essential that the excess of B required to complete the reaction is added to A. This end-point rounding and extrapolation is shown in Fig. 1.2.

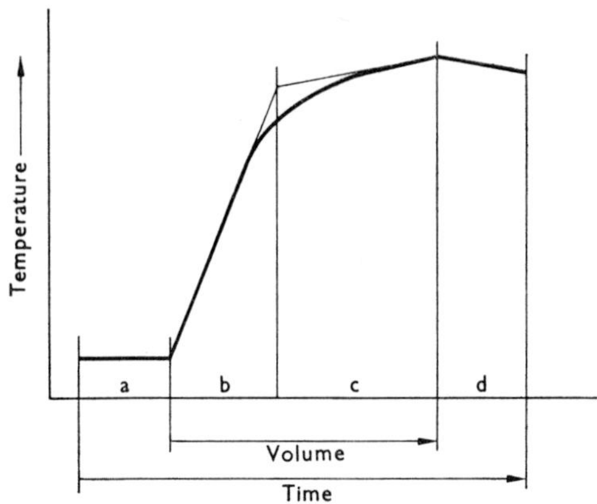

FIG. 1.2 Extrapolations in a rounded end-point thermometric titration.
(a) base period—no titrant added;
(b) reaction period—titrant added;
(c) excess titrant period—titrant added;
(d) post-titration period—no titrant added.

Tyrrell (1967) showed that the curvature, ξ, of the titration curve at the equivalence point could be expressed by the following equation:

$$\xi = 1/Kc_{A'} = (1 - \alpha)^2/\alpha \tag{8}$$

in which $c_{A'}$ is the concentration of A calculated as if it had been diluted by the small volume of B added at the equivalence point and α is the proportion of A that has been converted into P.

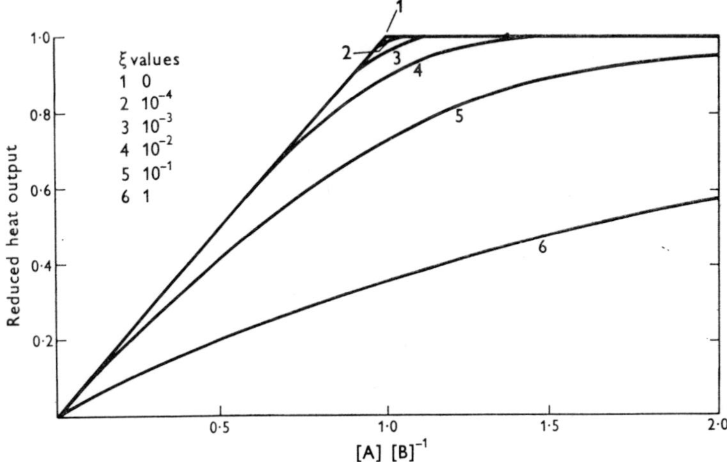

FIG. 1.3 Effect of the value of ξ on end-point sharpness 'reduced heat' titration curves; from Tyrrell (1967).

Thus if $\alpha = 99\%$ then $\xi = 10^{-4}$ and a sharp end-point will be obtained as shown in Fig. 1.3. The figure shows the calculated effects of different values of ξ on the shape of the titration curve and it can be seen that reasonably sharp end-points can be obtained with a value of ξ of 10^{-3}, i.e., from Eq. 8 $\alpha \approx 96.9\%$ and $Kc_{A'} \geqslant 10^3$. If, for example, a neutralization reaction of a 0.1 M solution of a weak acid is considered and if the acid has a dissociation constant, K_a, of 10^{-10} then $K = K_a/K_{H_2O} = 10^4$, and $Kc_{A'}$ has a value of 10^3 (K_{H_2O} is the ionic product for water, 10^{-14}). Therefore, under these conditions a sharp end-point is obtained and this is confirmed in the neutralization titration of 0.1 M phenol ($K_a = 1.28 \times 10^{-10}$) as is shown in Fig. 1.4.

At lower concentrations of an acid with a similar K_a-value the end-point is more curved, as shown in the titration of 0.01 M phenol in Fig. 1.4, and

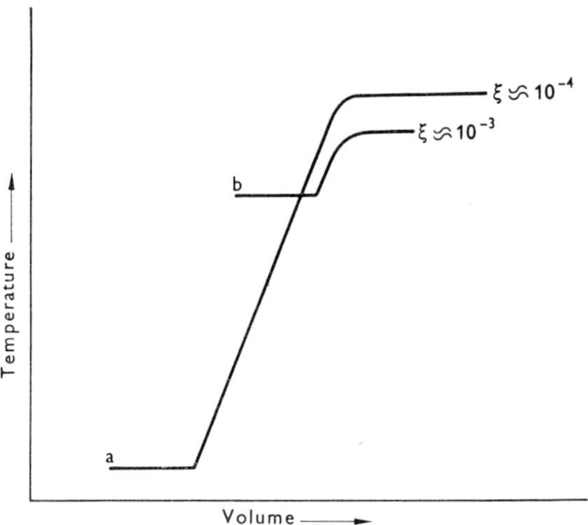

FIG. 1.4 End-point sharpness in the titration of phenol titration of (a) 10 ml of 0.1 M, (b) 20 ml of 0.01 M phenol with 5 M NaOH; from Vaughan (1970).

errors arising because of this can be reduced by the tangential extrapolation technique described above, but this requires a certain linear length of titration curve in the excess titrant period (c in Fig. 1.2). Jordan and Dumbaugh (1959b) suggested that this length should represent a 100% excess of titrant and that under these circumstances an end-point precision of $\pm 1\%$ can be obtained with acid concentrations of greater than 3 mM (i.e., $Kc_{A'} = 3 \times 10^{-1}$). Jordan (1968) stated that in the neutralization titration of the third ionization stage of 0.01 M H_3PO_4 ($pK_a = 12.36$) the concentration of sodium hydroxide required to permit an extrapolation to give an end-point precision of $\pm 1\%$ would exceed the solubility of sodium hydroxide in the system.

1.3.3 *Temperature of the system*

Equations 1 and 6 can be combined with the elimination of ΔG and referred to standard states to give Eq. 9

$$\frac{\Delta H}{T} = -R \ln K + \Delta S \qquad (9)$$

Thus, provided ΔH and ΔS are constant, the only effect of the temperature of the medium on thermometric or enthalpimetric titrations is that when the reaction is exothermic (ΔH negative), K increases as T decreases so that a lower temperature favours the completion of the reaction. Conversely, when the reaction is endothermic (ΔH positive), K increases as T increases so that a higher temperature favours the completion of the reaction.

The majority of titrations have been carried out at ambient temperatures with the notable exception of the work by Jordan and co-workers (1959b, 1960b) on exothermic reactions in molten solutions of salts but there is no reason why titrations should not be carried out at other temperatures and particularly at low temperatures, providing the kinetics of the reaction are favourable.

1.4 Kinetics of the reaction and the titration rate

1.4.1 *Titration rate*

The kinetics of the reaction being utilized or studied in the titration are important because in a thermometric titration the rate of the reaction must be greater than the rate of addition of the titrant, otherwise irregular titration curves will be obtained. In enthalpimetric titrations the temperature pulse will not be sharp, but under controlled conditions the titration curve in this case can be used to determine the rate of slow chemical reactions in solution (see Section 1.6), and in continuous-flow enthalpimetry, if the flow of the titrant and reactant is fast enough, the rate of fast chemical reactions can be determined (see Section 1.7.1).

In rapid ionic reactions such as neutralization, it has been estimated (Jordan and Dumbaugh, 1959b) that under the usual titration conditions, a titration rate of 0·02 ml/s is about a millionth of the rate of reaction so that the rapid titration rate of 10 ml/s used by Priestley (1963a) is still within the reaction rate. If in the much slower organic reactions the titration rate is too slow for practical purposes, it may be possible to use a catalyst to speed the reaction. The use of perchloric acid as a catalyst in the enthalpimetric titration of alkyl phenols with acetic anhydride (Snelson *et al.*, 1967) is an example of this.

1.4.2 *Use of kinetics in the titration*

The utilization of a fast reaction in the presence of a slow interfering reaction is one way in which kinetics can be used in the titration. For example, Jordan and Billingham, in their titration of calcium in the presence of magnesium, showed that the titration of magnesium with oxalate gave an isothermal reaction caused by the delay in the precipitation of magnesium oxalate, and that as a result, if calcium were present, it could be titrated exothermally during this period (see Section 3.6.2).

A second way in which kinetics have been used in a titration is in the 'catalytic indicator' method developed by Vaughan and Swithenbank (1965). They showed that if a substance or solvent were present in or added to the titrand and if this reacted with the titrant only after all the desired reaction had been completed, the end-point of the desired reaction was marked by a temperature change caused by the reaction with the added substance. Even when the desired reaction produces only a small temperature change the relatively larger amount of the substance that can be added ensures a large temperature change at the end-point. Under these conditions the rate of addition of the titrant is important, as shown in Table 1.2. The more rapid the titration the greater is the possibility of positive error. This is caused by the titrant concentration not being reduced quickly enough to a value below the threshold limit of its catalytic effect.

Table 1.2 Effect of titrant rate, using the catalytic indicator method. Titration of silver with 0.01 M KI (Ce(IV)–As(III) system)
(From Weisz *et al.* (1969))

Titration rate, μl/min	Relative error, %
5	$-1{\cdot}00$
55	$+0{\cdot}95$
556	$+1{\cdot}7$
5568	$+3{\cdot}7$

Details of the methods used in the determination of acids or bases in non-aqueous solution are given in Sections 4.1.2 and 4.1.4(*i*) respectively and of cations and anions in aqueous solution in Section 3.52.

The 'kinetic titration' described by Sajó (1968a) in which the velocity of a catalysed reaction is utilized in a differential system is a third way in which kinetics have been used. In a homogeneous catalytic reaction, provided that the concentrations of the reactants remain almost constant and the temperature change is small, then the rate of temperature change will be proportional to the concentration of the catalyst $[C]$ and to the velocity of the reaction dc/dt, i.e.,

$$dT/dt = k[C]\, dc/dt \tag{10}$$

When a substance, S, to be determined behaves as a catalyst for a particular reaction and two solutions are prepared alike in every respect except that one contains the substance S in a sample, and the reactants, in high concentration, are injected into both solutions simultaneously, then the temperature of the sample solution will rise in proportion to the concentration of substance S in it. If substance S is then titrated into the 'reference' solution containing no sample until its temperature is the same as that of the sample solution, then the quantity of substance S added is the same as that in the sample (see also Section 3.37.2).

Note

Papoff and Zambonin (1967) investigated the use of thermometric titrations in the study of chemical kinetics. They used simple apparatus that required no calibration in the calculation of the kinetic parameters of moderately fast and of consecutive reactions. The method indicates that relatively slow reactions may be used in analysis (see Section 4.6).

1.5 Fundamentals of a thermometric titration

The number of moles of product formed, n_P, is related to the volume of titrant added, V_t, and to the molarity of the titrant, M_t, by the equation $n_P = M_t V_t$ and therefore substituting for n_P in Eq. 5 gives:

$$\Delta T = -\Delta H M_t V_t / C_s \tag{11}$$

The total heat capacity of the system, C_s, consists, however, of two parts, that of the medium, C_m, at the start of titration and that of the apparatus, C_a, comprising the container, stirrer etc. Thus

$$C_s = C_m + C_a \tag{12}$$

The medium consists of the initial volume of the titrand, V_0, plus the volume of added titrant, V_t, therefore:

$$C_s = V_0 E_0 + V_t E_t + C_a \tag{13}$$

where E is the volume specific heat, i.e., $C = \rho E V$, where ρ is the density of the fluid.

Combining Eqs. 11 and 13 to eliminate C_s,

$$\Delta T = \frac{-\Delta H\, M_t V_t}{V_0 E_0 + V_t E_t + C_a} \tag{14}$$

The titrant has a different temperature from that of the medium during the titration even if they were the same at the beginning. If this difference between the temperature of the titrant, T_t, and that of the initial temperature of the medium (titrand), T_0 is $T_t - T_0$, then the actual temperature change $\Delta T'$ that will be obtained is given by Eq. 15.

$$\Delta T' = \frac{-\Delta H\, M_t V_t}{V_0 E_0 + V_t E_t + C_a} + \frac{V_t E_t (T_t - T_0)}{V_0 E_0 + V_t E_t + C_a} \tag{15}$$

$$\therefore\ \Delta T' = [-\Delta H\, M_t + E_t\,(T_t - T_0)] \frac{V_t}{V_0 E_0 + C_a + V_t E_t}$$

Differentiating $\Delta T'$ with respect to V_t gives the slope of the titration curve:

$$\frac{d\Delta T'}{dV_t} = [-\Delta H\, M_t + E_t(T_t - T_0)] \frac{V_0 E_0 + C_a}{(V_0 E_0 + C_a + V_t E_t)^2} \tag{16}$$

1.5.1 *Volume of the titrant*

Equation 16 shows that unless V_t, the volume of the titrant, is small, the titration curve will not be linear during the reaction period, giving a less pronounced end-point. However, linearity can be achieved by increasing the molarity of the titrant, i.e., reducing the quantity of titrant added to obtain the same quantity of product. Ideally this should be done by generating the titrant within the titrand, i.e., zero volume of titrant. This has been achieved by Vajgand and his co-workers (1970) by the use of a coulometrically generated titrant, an advance which should have a very wide application. If an external titrant is used it is usually between 20 and 1000 times the molarity of that of the titrand (*cf.* Ewing and Mazac, 1966). If the titrant volume is large, i.e., if the titrant concentration is low, the effect on the curvature of the titration curve may be minimized by replotting the curve, using the correction factor $V_c = V_0 + V_t/V_0$ (Chatterji, 1958b).

If the volume of titrant used is small, Eq. 16 reduces to the form

$$\frac{d\Delta T'}{dV_t} = \frac{-\Delta H \, M_t + E_t(T_t - T_0)}{V_0 E_0 + C_a} \tag{17}$$

1.5.2 *Temperature of the titrant*

Equation 17 shows that the slope of an exothermic titration curve, when compared with one where the titrant is initially at the same temperature as the titrand ($T_t = T_0$), will be steeper if $T_t > T_0$ and less steep if $T_t < T_0$. When the reaction has been completed the titrant will then bring the temperature of the titrand exponentially to its own temperature. Thus if the titrant temperature is less than the final temperature, the post-reaction titration curve will show an endothermic portion that will be initially a

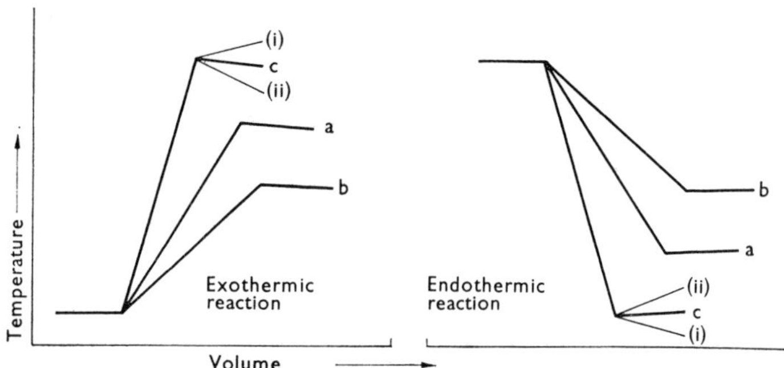

FIG. 1.5 Effect of titration temperature on a thermometric titration curve.
    a titrant and titrand initially at the same temperature
    b titrant initially cooler than the titrand;
    c titrant initially warmer than the titrand;
        (*i*) titrant warmer than the final solution
        (*ii*) titrant cooler than the final solution.

straight line with negative slope, the slope being greater the lower the temperature of the titrant, compared to the final temperature. If the titrant is warmer than the final solution the post-reaction titration curve will show a linear exothermic portion. Similar arguments apply to endothermic reactions and both effects are shown in Fig. 1.5.

Though the effects of the titrant temperature are not necessarily large it can be seen that the end-point can be sharpened for an exothermic reaction by the use of a lower-temperature titrant (Jordan and Dumbaugh, 1959a) and for an endothermic reaction by a higher-temperature titrant. A practical demonstration of the effect for an exothermic reaction is shown in Fig. 3.4.

1.5.3 *Significance of the initial slope*

If the titrant and titrand have the same temperature at the start of the titration (and this is often the case as the burette tip is in contact with the titrand during the base period), then because $T_t = T_0$, Eq. 17 becomes:

$$\frac{d\Delta T}{dV_t}(V_t = 0) = \frac{-\Delta H\, M_t}{V_0 E_0 + C_a} \tag{18}$$

The initial slope of the titration curve is therefore proportional to $-\Delta H$ and thus the value of ΔH can be determined accurately from it (Keily and Hume, 1956), provided that the value of $V_0 E_0 + C_a$, which equals the initial total heat capacity, C_s, (*cf.* Eq. 12) can be determined. This total heat capacity can be determined by a calibration procedure carried out under similar conditions (see Section 2.5) or, if the heat of a reaction is known, a titration using the reaction can be carried out and the value of C_s determined by utilizing the value of the initial slope so obtained (*cf.* Mead, 1962).

The initial slope method, which will give a value of ΔH that approaches that of the enthalpy at infinite dilution, $\Delta H°$, cannot be used if the kinetics during the initial stages of the reaction are slow because of the small quantity of titrant added. Under these circumstances the graphical extrapolation method can be used (see Section 1.5.5) because the extrapolation is made from those parts of the titration curve at which the process has reached completion.

1.5.4 *Titrand volume and heat capacities*

It can be seen from Eq. 18 that the slope of the titration curve during the reaction period is increased and thus the sharpness at the end-point is improved, if the volume, V_0, and the volume specific heat, E_0, of the titrand are decreased. The same effect is obtained by reducing the heat capacity of the apparatus, C_a.

Another effect of a low specific heat of the titrand, as found for example in organic solutions, is the greater temperature rise for a particular heat change, i.e., a greater sensitivity can be obtained.

1.5.5 *Characteristics of the titration curve*

It has been shown that the thermometric titration curve is a function of both the enthalpy change, ΔH, and the equilibrium constant, K, of the reaction. If therefore the reaction between B and A to produce P has proceeded some way by the addition of B to the solution containing A, then the total concentrations of B, $[B_t]$, and A, $[A_t]$ are given by the following;

$$[A_t] = [B] + [P] \tag{19}$$

$$[B_t] = [A] + [P] \tag{20}$$

But $[A]$, $[B]$ and $[P]$ in the solution are related to the equilibrium constant by Eq. 7.

$$K = \frac{[P]}{[A][B]} \times \frac{\nu_P}{\nu_A \nu_B} \tag{7}$$

From Eq. 3, if V is the volume of the solution

$$Q = -\Delta H[P]V \tag{21}$$

and therefore combining Eqs. 19, 20, 7 and 21 with the elimination of $[A]$, $[B]$ and $[P]$ the following general equation is obtained:

$$-\frac{\Delta H}{K} = a(\Delta H)^2 + b\,\Delta H + c \tag{22}$$

where

$$a = V[A_t][B_t]\nu_A\nu_B/Q\nu_P$$
$$b = ([A_t] + [B_t])\nu_A\nu_B/\nu_P$$
$$c = Q\nu_A\nu_B/V\nu_P$$

In Eq. 22 the ν values are a function of the ionic strength, μ, which in its turn is a function of K, and thus Eq. 22 contains only the two unknowns ΔH and K. Because $\Delta H/K$ has the same value at any point on the titration curve, Eq. 22 can be solved for ΔH by the use of two points on the titration curve according to the following equation:

$$(a_2 - a_1)(\Delta H)^2 + (b_2 - b_1)\,\Delta H + (c_2 - c_1) = 0 \tag{23}$$

The value of ΔH having been obtained by the use of Eq. 23, K (and thus the free energy change ΔG by Eq. 6) can be calculated by using Eq. 22, and finally the entropy change, ΔS, can be calculated by using Eq. 1.

The method of calculation of the thermodynamic values, ΔH, ΔG and ΔS was devised by Hanson, Christensen and Izatt (1965), but in order to obtain self-consistency between K, ν and μ values, the value of K obtained was used to calculate the concentrations of the components present. These in their turn were then used to calculate new ν and μ values, which were then used to calculate a better value of K. However, this method is subject to criticism (see page 228). A computer was used to repeat this calculation until successive values of K agreed within 0.1%. Wanders and Zweitering (1969) confirmed that this iterative technique fully utilized the high precision obtainable with the apparatus developed by Christensen et al. (see Section 2.2.1(i)), and Holmes and Williams (1967a) also used computer techniques to calculate ΔH from thermometric data.

Similar arguments can be applied to multicomponent or multi-ionization reactions of the form:

$$A + B \rightleftharpoons AB$$
$$A + 2B \rightleftharpoons AB_2$$
$$\cdot \quad \cdot \quad \cdot \quad \cdot \quad \cdot \quad \cdot$$
$$A + nB \rightleftharpoons AB_n$$

The equations equivalent to Eqs. 19, 20, 7 and 21 above are then respectively:

$$[A_t] = [A] + [AB] + 2[AB_2] \ldots + n[AB_n] \tag{24}$$
$$[B_t] = [B] + [AB] + [AB_2] \ldots [AB_n] \tag{25}$$

$$K_n = \frac{[AB_n]}{[A][B]^n} \times \frac{\nu_{AB_n}}{\nu_A(\nu_B)^n} \tag{26}$$

$$Q = V \int_1^n [AB_n][\Delta H_n] \tag{27}$$

When Eqs. 24–27 are combined, the single equation equivalent to Eq. 22 will contain $2n$ unknowns—n K-values and n ΔH-values. If then values of K_n are assumed and the corresponding concentrations of the complex stages or ionizations are calculated for each of $2n$ thermometric measurements, the repetitive computer calculation described above can be used (Izatt *et al.*, 1968).

If ΔH is to be determined from a thermometric titration curve the value obtained from the rise in temperature in the reaction period (b, Fig. 1.1a) is only approximate. This error caused by titrant–titrand temperature differences and heats of dilution can be minimized by extrapolation of the curve back from the excess-titrant and post-titration periods of the titration curve (c and d, Fig. 1.1a).

This extrapolation method, which is described in Section 2.6.2 and also shown in Fig. 2.32 has been the subject of a theoretical evaluation by Carr (1972) who states that the accuracy of the method is not only dependent on the details of the extrapolation procedure but also on the magnitude of the heat transfer characteristics of the calorimeter and the rate of change of the heat capacity caused by the titrant addition. He proposed the use of a simple mathematical algorithm which led to low errors, of about $0.1–0.5\%$.

1.5.6 *Multicomponent titrations*

The discussions in this section so far have been concerned with single-component titrations but, as in potentiometric titrations, under certain conditions, components of a multicomponent system or the products of individual dissociation stages of a polyprotic molecule can be titrated serially in a single titration. The ability to carry out a serial thermometric analysis depends on two factors. The first is that the titration of one component or stage should be virtually complete before that of the second component or stage commences, and this will obviously depend on the values of the equilibrium constants of the two reactions. Kortüm and co-workers (1961) stated that the ratio of the two equilibrium constants K_1/K_2 should not be less than 10^3, i.e., $pK_2 - pK_1$ should not be less than 3, otherwise there is an appreciable overlap of the dissociations. In making this statement Kortüm *et al.* were not considering thermometric studies, but these more than support it in that it is possible to distinguish acids having pK_a differences as low as 2 (see Section 3.48) or even 1.45 (see Section 4.1.1(*iii*)). The second factor is that there should be a sufficient difference between the two heats of reactions, i.e., their respective titration slopes should be sufficiently different to produce a clear inflection between them. It has been stated by Forman and Hume (1964) that a slope ratio of $3:2$ is necessary for the inflection to be clearly defined. Figure 1.6

shows the titration curve of an analysis of this type in which the free acid and three base sulphates are titrated serially with sodium hydroxide.

Ewing (1961) stated that a serial analysis can be carried out if successive free energies and heats of reaction differ by 3–5 kcal/eq.

A completely different approach to this problem has been put forward

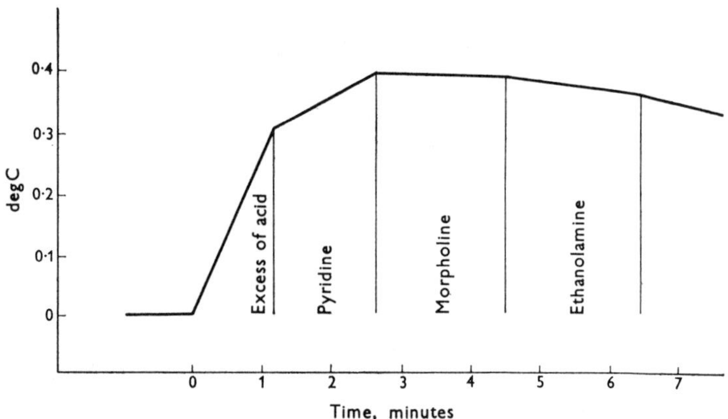

FIG. 1.6 Titration curve of an aqueous solution of pyridine, morpholine and ethanolamine, slightly acidified with sulphuric acid, with 5 M NaOH; from Vaughan and Swithenbank (1967).

by Hansen and Lewis (1971a) who considered the titration of a binary mixture of reactants, A and B, which had equal or nearly equal equilibrium constants. The total number of moles, n_T, is then the sum of the number of moles of reactants, n_A and n_B.

$$n_T = n_A + n_B \tag{28}$$

The total heat evolved, Q_T, is also the sum of the heats of the individual reactions, Q_A and Q_B.

$$Q_T = Q_A + Q_B \tag{29}$$

Substituting for Q_A and Q_B by using Eq. 3,

$$Q_T = n_A \Delta H_A + n_B \Delta H_B \tag{30}$$

If Eqs. 28 and 30 are then combined, with the elimination of n, and rearranged in terms of n_A,

$$n_A = \frac{Q_T - n_T \Delta H_B}{\Delta H_A - \Delta H_B} \tag{31}$$

The total heat evolved, Q_T, and the total number of moles reacting, n_T, can be calculated from a single thermometric titration curve. Therefore knowing these and the heats of reaction ΔH_A and ΔH_B of the two reactants, it is possible to calculate the number of moles of A (and of B) by using Eq. 31. The equation also shows that the smaller the difference between ΔH_A and ΔH_B the less sensitive the method.

The precision of the method was found to be $\pm 5\%$ at the mM level of each component in the test systems used, but was critically dependent on the sharpness of the end-point and also required the use of precise apparatus. It will obviously be of use in the analysis of bifunctional molecules, but the need to know the enthalpies of each reaction is a limitation, particularly as published or determined values of these may be different from those under the conditions used in the titration.

The authors stated that this 'total moles–total enthalpy' method could be extended to a ternary system provided that the total mass of the reactants was known.

1.5.7 *Adiabaticity of the system*

It has so far been assumed that the heat of reaction takes place in a system from which there is no loss or gain of heat. Any loss or gain of heat can be minimized by proper design of the calorimeter, but where precise measurements of the heat of reaction are to be made such effects can be overcome in a number of ways. If it is assumed that the environment is at the same temperature as the titrand there is no heat movement but if the titration is, for example, an exothermic one, the heat loss, though small, will be greater the longer the titration, and the total temperature rise will be less than that expected from the reaction.

Schlyter and Sillén (1959) suggested that the temperature of the titrand

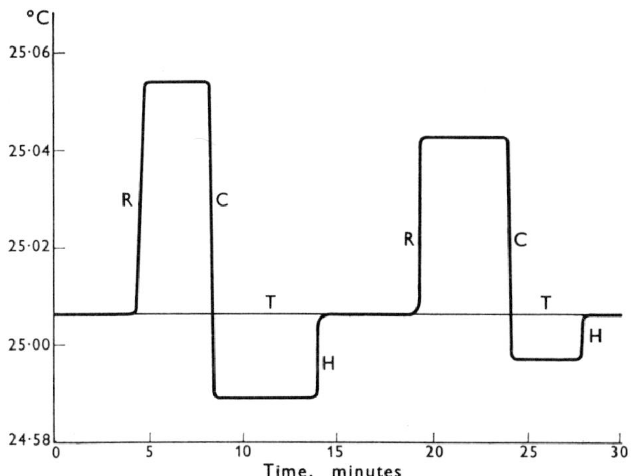

FIG. 1.7 Exothermic titration curve with temperature adjustment; from Schlyter and Sillén (1959).
Temperature changes: R, from reaction; C, from cooler; H, from heater. Thermostat temperature, T.

after each incremental addition of the titrant should be brought back to the original temperature by the use of a cooler or heater as appropriate so that, because the maximum difference in temperature between the titrand and its environment was then of the order of 0·05 degC, the heat gain or loss would be negligible.

Figure 1.7 shows a typical exothermic titration curve in which it can be seen that the titrand was overcooled and then heated to bring it to the thermostat temperature; this was because greater control of the temperature could be obtained by using the heater.

Christensen, Johnston and Izatt (1969) used a different principle involving an isothermal calorimeter and continuous titration. In this method the titrand was maintained at a constant temperature by extracting heat from it by the Peltier effect and balancing this by adding heat from a variable-energy heater. Thus any heat changes caused by the titration are shown by the change in the heat output of the variable heater required to keep the temperature of the titrant constant.

If the titration takes place extremely rapidly the thermal inertia of the calorimeter is such that heat exchange with the environment is negligible during the period of the titration. This is the basis of the rapid titration method of Priestley, Sebborn and Selman (1963) who showed that under these conditions only extremely simple apparatus was required.

Nakanishi and Fujieda (1972) considered a differential titration system in which the heat evolved was allowed to equilibrate freely with the equal-temperature surroundings of the two simple reaction vessels. They developed the following equation to describe these non-adiabatic conditions.

$$\frac{1}{C_s}\int q\,\mathrm{d}t = \frac{1}{\beta}\frac{\mathrm{d}\theta}{\mathrm{d}t} + \left(1 + \frac{\alpha}{\beta}\right)\theta + \alpha\int\theta\,\mathrm{d}t \tag{32}$$

where t = time in s

q = heat evolved in cal/s

C_s = effective heat capacity in cal/degC

θ = temperature difference between the two calorimeters

α = a constant (s^{-1}) related to heat transfer

β = a constant (s^{-1}) related to delay of response.

The left-hand side of Eq. 32 is equal to the temperature change of the sample solution under perfect adiabatic conditions and thus can be calculated from the known values of θ under non-adiabatic conditions. The authors found it preferable to feed the voltage from the Wheatstone bridge that represented θ into an operational amplifier, the characteristics of which could be matched to represent Eq. 32. In this way, although the end-points in the titration of $0{\cdot}1$–$1{\cdot}0$ mmole of phenol in 50 ml of solution were slightly rounded, the error was less than $\pm 0{\cdot}7\%$.

1.6 The fundamentals of an enthalpimetric titration

If the equilibrium reaction of the type given in Eq. 2

$$aA + bB \rightleftharpoons pP \tag{2}$$

is considered when an excess of B has been added to A so that the reaction can go to completion, then provided the reaction velocity is greater than the rate of addition of B the rapid temperature rise of an enthalpimetric titration will be obtained. If however the reaction velocity is slower than the titration rate, the temperature rise will not be rapid and under these circumstances the velocity of slow reactions can be calculated from the temperature–time relationship obtained as described by Meites *et al.* (1969).

If the reaction velocity is fast and the formation of the product, P, is virtually complete, then under these circumstances, using Eq. 2,

$$n_P = \frac{p}{a} n_A \tag{33}$$

where n_A is the number of moles of A (the unknown) that are present.

According to Eq. 5, however,

$$\Delta T = -\Delta H \, n_P / C_s \tag{5}$$

Therefore, combining Eqs. 5 and 33 with the elimination of n_P

$$\Delta T = -\Delta H \, n_A p / a C_s$$

or

$$n_A = -\Delta H \, C_s a / p \, \Delta H \tag{34}$$

If C_s, the heat capacity of the system, is constant, which can be virtually achieved by adding B as a concentrated solution (see also Section 1.5.1), and because the values of ΔH, a and p for a particular reaction are all constant then:

$$n_A = k_1 \, \Delta T \tag{35}$$

The relationship between the concentration of A, c_A, and the temperature change can be obtained by dividing Eq. 35 by the volume of the test solution, V_0, to give

$$c_A = \frac{n_A}{V_0} = \frac{k_1 \, \Delta T}{V_0} \tag{36}$$

If, therefore, each determination of A is always carried out under the same volume conditions,

$$c_A = k_2 \, \Delta T \tag{37}$$

or

$$A\% = k_3 \, \Delta T \tag{38}$$

The constants k_1, k_2 and k_3 in Eqs. 36, 37 and 38 can be determined by calibrating with known quantities of A, but the calibration obtained can only be used in the determination of A under the conditions of the calibration. The arguments above apply equally to the normal addition of excess of B to A as to the addition of the sample A to excess of B (*cf.* Snelson *et al.*, 1967).

Because an enthalpimetric titration is concerned with the overall reaction it cannot be used, as is a thermometric titration, to distinguish between components of a multicomponent mixture or multi-ionization stages.

1.7 Fundamentals of continuous-flow enthalpimetry

In the discussions above it was assumed that the titrant was added to a static titrand, but it can be advantageous to have both titrant and titrand as continuous flows. This technique has been used to measure reaction velocities and heats of mixing and in continuous analysis.

1.7.1 *Measurement of reaction velocity*

If constant flows of a titrant containing A and of a titrand containing B, each at the same temperature and having the same thermal capacity, are mixed together to form a product P and if the mixture flows through a narrow tube of known diameter and the temperature of the product solution is measured at known distances along the tube from the mixing point then the temperature rise, when plotted against the time taken to

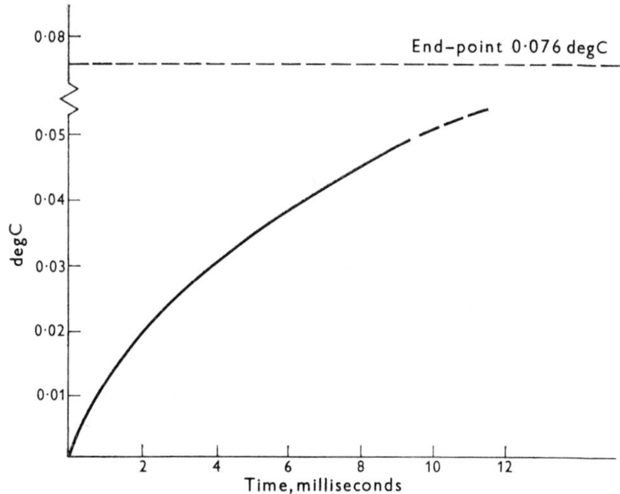

FIG. 1.8 Enthalpimetric measurement of rate of reaction; from Roughton (1963). Reactants: 0·00735 M CO_2 and 0·404 M NaOH at 20 °C.

reach the measuring point as shown in Fig. 1.8, will give a curve from which the rate of combination of A with B can be obtained.

This is the basis of the work of Roughton who measured velocity constants of between 80 and 24000 l $mole^{-1} s^{-1}$ of a wide variety of reactions. Those interested in this work should read his review (Roughton, 1963).

1.7.2 *Continuous analysis*

If the mixture above flows through a wider tube, i.e., sufficient time is allowed for the reaction to be completed, then the temperature rise will be a measure of A or B, whichever is the lesser, or alternatively, it will be a measure of the heat of mixing, if the two solutions do not contain reactants (see Section 1.8.3).

Under reaction conditions it follows that if A in one flow is always in excess of B in the other flow, the temperature rise will be a measure of B. This is the basis of the continuous enthalpimetric method developed by Priestley and his co-workers (1965), which can be used for continuous analysis and on-line control of processes.

Assuming that the flow-rate (ml/s) of the sample flow containing B is f and that of the reactant A is Rf, i.e., their flow-rate ratio is R, then the molar flow-rates in mmole/s will be $M_B f$ and $M_A Rf$ respectively, where $M_A Rf$ is always greater than $M_B f$.

If the heat of the reaction between A and B is H kcal/mole and the temperatures of the flows containing A, B and P are T_A, T_B, and T_P respectively, then the heat balance is as follows.

(a) Heat entering the system

(i) due to the reaction enthalpy

$$fM_BH$$

(ii) due to the unreacted solutions

$$C_BfT_B + C_ARfT_A$$

(b) Heat leaving the system

(i) due to the product solution

$$C_P(R + 1)fT_P$$

Thus at equilibrium when the net heat change is zero

$$fM_BH + C_BfT_B + C_ARfT_A = C_P(R + 1)fT_P \tag{39}$$

Assuming that the volume heat capacities C_B, C_A and C_P are similar, as will be so if the solutions have the same solvent and their concentrations are less than 1 M, then Eq. 39, rearranged, becomes

$$\frac{M_BH}{C} = T_P(R + 1) - T_B - RT_A \tag{40}$$

But in a particular reaction H and C will be constant, i.e., $C/H = k_1$ and therefore

$$M_B = k_1[T_P(R + 1) - T_B - RT_A] \tag{41}$$

Thus the molarity of the flow containing the sample, M_B, is proportional to the three solution temperatures, T_A, T_B and T_P, and also to the flow-rate ratio, R.

The flow-rate term, f, is absent from Eq. 41 and therefore flow-rates are unimportant provided the flow-rate ratio, R, is constant, though fluctuation of incoming flow-rates is to be avoided because of its effect on the equilibrium.

If the sample and reactant flow temperatures are the same, i.e., $T_B = T_A = T_R$, then Eq. 41 becomes

$$M_B = k_1(T_P - T_R)(R + 1) \tag{42}$$

Therefore, if R is kept constant,

$$M_B = k_2(T_P - T_R) \tag{43}$$

Under these circumstances, the sample molarity M_B is proportional to the difference between the temperature of the product flow and that of the incoming flows.

1.8 Heats of solution, dilution and mixing

It was assumed in the previous discussion that on addition of the titrant there were no effects caused by the heat of solution, dilution or mixing. This is obviously not necessarily the case, particularly as fairly concentrated titrants have been shown to be essential in both thermometric and enthalpimetric titrations. These extraneous heat effects must be kept within reasonable limits, and Jordan (1968) suggested that their total heat-effect in the titration should not exceed one third of the heat of the reaction being studied or utilized. Ways of reducing or overcoming these effects where they interfere are discussed later, but heats of solution, dilution and mixing have all been determined by thermometric titrimetry. The heat of dilution has also been utilized in end-point detection.

1.8.1 *Determination of the heat of solution*

Keily and Hume (1956) showed that the slope of a thermometric titration curve, dT/dV_t, could be used to determine the heat of solution. When a liquid solution is added to a solvent the following relationship is applicable:

$$\frac{dT}{dV_t} = -\frac{(\bar{L}_2 - L_2^*)M}{C} \tag{44}$$

where $(\bar{L}_2 - L_2^*)$ is the differential heat of solution.

At the beginning of the titration the relative partial molal heat content of the solute $\bar{L}_2$ is zero so that Eq. 44 becomes

$$\frac{dT}{dV_t}(V_t = 0) = \frac{L_2^* M}{C_s} \tag{45}$$

where L_2^* is the relative molal heat content. The initial slope of the thermometric titration curve therefore provides an easy means of determining the differential heat of solution in infinitely dilute solution.

In the titration of the solvent into a pure liquid solute the following equation can be used:

$$\frac{dT}{dV_t} = -\frac{\bar{L}_1 M}{C_s} \tag{46}$$

In this case, $\bar{L}_1$, the relative partial molal heat content of the solvent, is not constant but varies with the composition and approaches zero as more solvent is added.

1.8.2 *Determination of the heat of dilution*

The heat change on titration of a pure solvent into a solution is analogous to the titration into a pure liquid solute and Eq. 46 applies. Similarly if the pure liquid solute is titrated into the solution, Eq. 45 applies. However, when a solution is titrated into a pure solvent the following relationship is applicable.

$$\frac{dT}{dV_t} = -\frac{(\bar{L}_1^i - \bar{L}_1^f)M_1 + (\bar{L}_2^f - \bar{L}_2^i)M_2}{C_s} \tag{47}$$

where the superscripts i and f refer to the initial and final states respectively, and the subscripts 1 and 2 refer to the solvent and solution respectively. At the start of the titration $\bar{L}_1^f$ and $\bar{L}_2^f$ both have zero value and Eq. 47 becomes

$$\frac{dT}{dV_t}(V_t = 0) = \frac{\bar{L}_1^i M_1 + \bar{L}_2^i M_2}{C_s} \tag{48}$$

Thus if either $\bar{L}_1$ or $\bar{L}_2$ (relative partial molal heat contents) is known the other can be calculated from the value of the initial slope of the titration curve. In the more common situation where one solution is added to another, Eq. 42 is applicable.

1.8.3 *Determination of the heat of mixing*

The molar excess energy of mixing at constant volume, ΔU_v, according to the solubility parameter theory of Scatchard and Hildebrand (Hildebrand and Scott, 1962), is given by Eq. 49 in which the subscripts 1 and 2 refer to the two components being mixed.

$$\Delta U_v = (x_1 V_1 + x_2 V_2)(\delta_1 - \delta_2)^2 \phi_1 \phi_2 \tag{49}$$

where x is the mole fraction, V is the partial molar volume, δ is the solubility characteristic, ϕ is the volume fraction of the component. The factor $(x_1 V_1 + x_2 V_2)$ is the molar volume of the mixture, so if ΔQ_v is the excess energy of mixing at constant volume divided by the total molar volume, Eq. 49 becomes

$$\Delta Q_v = (\delta_1 - \delta_2)^2 \phi_1 \phi_2 \tag{50}$$

Therefore, provided $(\delta_1 - \delta_2)$ is a constant (and this is generally the case)

$$\frac{\Delta Q}{\phi_1 \phi_2} = k \tag{51}$$

Sturtevant and Lyons (1969), who have used continuous flow enthalpimetry to determine heats of mixing, suggested that values of the expression $\Delta Q V / \phi_1 \phi_2$ should provide a more convenient and significant expression for the experimental results than that more usually employed, namely ΔH_m, the molar excess enthalpy of mixing.

1.8.4 *End-point enhancement by heat of dilution*

Consider a thermometric titration system in which a titrant containing A produces a large heat of dilution when it is titrated into a particular solvent, primarily because it contains A. Therefore when the solvent contains the reactant B and it is titrated with the titrant containing A, the resultant titration curve will show an initial temperature change caused by the reaction of A with B with no heat of dilution effects. When all of B has reacted, the further addition of titrant will produce a large temperature change because of the heat of dilution and this enhances what might have been a slight inflection at the end-point. This is the principle of end-point enhancement by heat of dilution, first proposed by Vaughan and Swithenbank (1967) in their determination of bases in non-aqueous solution (see Section 4.1.4).

1.9 Differential titrations

A differential thermometric titration is one in which two identical titration systems are used and the differential temperature changes are recorded during the titration in which an identical titrant is added at the same rate to each titrand. The technique can be used to negate heats of dilution or other extraneous heat effects during the titration (see Section 2.6.1) but it can also be used as a method of analysis.

1.9.1 *Dual titration analysis*

Taubinger (1969b) demonstrated that a type of differential titration can be used as a unique method of analysis in which the substance to be determined is used as the titrant. If one of the two identical titration vessels contains a sample in which the substance, S, to be determined is present and the other vessel contains a blank solution and to both of these solutions is added an equal quantity of a reactant, R, so that it is in slight excess of S, then, if the substance S is titrated into both solutions, the recorded differential temperature will show three distinct periods.

(*i*) No temperature change, as R in both solutions is titrated until the excess of R in the sample solution has been titrated.

(*ii*) A temperature change, as R in the blank solution is titrated, while there is no reaction in the sample solution.

(*iii*) No temperature change because both solutions are identical.

The amount of S added during the period of change of temperature is thus equivalent to the amount of S in the added sample. A typical titration

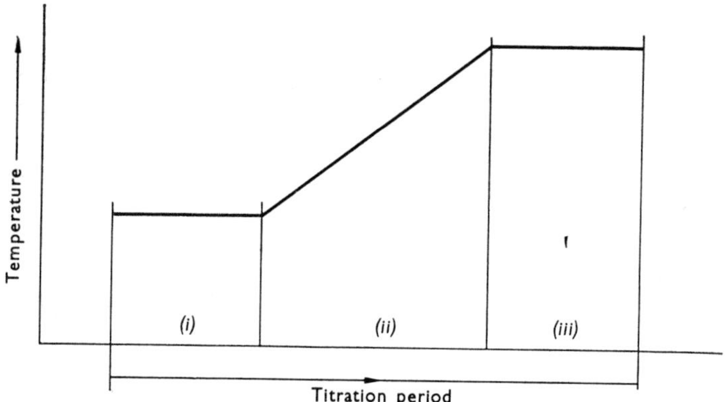

FIG. 1.9 Differential titration determination of vanadium(V) with vanadium(V) as titrant using hydrogen peroxide as reactant.
(*i*), (*ii*), (*iii*)—see text for explanation.

curve is shown in Fig. 1.9. The term dual titration has been used by the author to distinguish this type of titration from differential titrations used to negate heats of dilution etc.

Chapter 2

Practical Aspects

The apparatus required to carry out an analysis that is based on the conditions cited in the theoretical section is conveniently made up of three parts: (i) a titrant system for the addition or generation of the titrant, (ii) a titration vessel with temperature sensing device, stirrer, heater and cooler, and (iii) the device for measuring and/or recording the temperature. Because the analysis is based on calorimetry both the titrant system and the titration vessel must have some type of environmental control. Descriptions of the apparatus that has been used and its environmental control and of the final stage, the calculation and presentation of the results, are given in the following sections.

2.1 The titrant system

2.1.1 *Incremental addition*

Incremental addition from a thermostatted burette (see Fig. 2.3) was used in the early titration method because the Beckmann thermometer used had a large heat capacity and this necessitated a pause between each addition to allow thermal equilibrium to be established and the thermometer to be read, or alternatively the solution to be brought to a standard temperature before the next addition of the titrant. The burette was surrounded by a constant-temperature water-jacket, but a finer control of the titrant temperature was obtained by Schlyter and Sillén (1959) who used an apparatus in a room thermostatted at 25 ± 0.2 °C. The titrant at room temperature was adjusted to 25 ± 0.004 °C by passing it though a bulb immersed in the same water-thermostat as the titration vessel.

With the advent of low heat-capacity temperature-detection, continuous addition of the titrant became possible but incremental addition has certain advantages.

Goyan and co-workers (1961) used an automatic pipette with check valves, the titrant being delivered to the pipette by means of regular pulses of a mercury reservoir, and in this way 0·1 ml of titrant was delivered rapidly every 40 seconds. Incremental addition may be necessary in precise work to allow time for calorimeter equilibration (see Section 2.2.1(*i*)). For this type of work Gerding and his co-workers (1963) used the pipette shown in Fig. 2.1 which had a 2–5 ml bulb with a short exit-tube and contained a central glass capillary inner tube of 1 mm bore, sealed at its lower end, and ground to give a leak-proof joint in an internal seal lip above the exit tube. A small hole was drilled in the capillary tube just above its lower end. The pipette was connected to the inner tube by means of rubber tubing which had a tube connection to a blowing ball. The pipette was filled through the inner tube by means of a precision all-glass syringe and emptied by raising the inner tube, the last trace of titrant being displaced by operating the blowing ball. The bulb and exit-tube were immersed in the titrant to establish thermal equilibrium and then raised above the surface for the addition which, because all the inner surfaces had been treated with a silicone, gave a volume precise to within $\pm 0.04\%$ (see also Fig. 2.10).

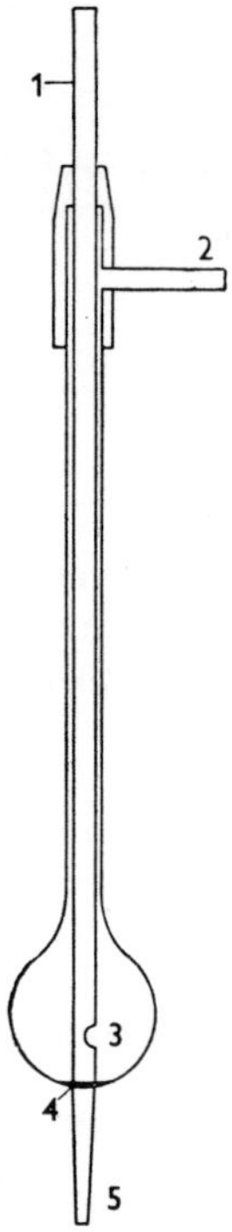

Brown and co-workers (1969) used a manually operated syringe burette, and after adding 90% of the required titrant continuously added small (about 0·1 ml) identical increments at regular intervals and measured the temperature after each addition, using a thermistor. They showed that the measurement could be satisfactorily plotted on a recorder chart by advancing the chart manually after each addition so that the pen crossed one of the regular time-indication lines on the chart. A typical titration graph is shown in Fig. 2.2 in which it can be seen that extrapolations can be made with much greater accuracy by joining the points where the temperature line crosses the time-indication line.

To operate this method it is necessary to know the approximate content, which is a disadvantage with a completely unknown sample in that a preliminary titration must be carried out to obtain an approximate end-point.

FIG. 2.1 Titration pipette; from Gerding *et al.* (1963). 1, capillary inner tube; 2, connection to blowing ball; 3, inlet hole; 4, ground seal; 5, exit tube.

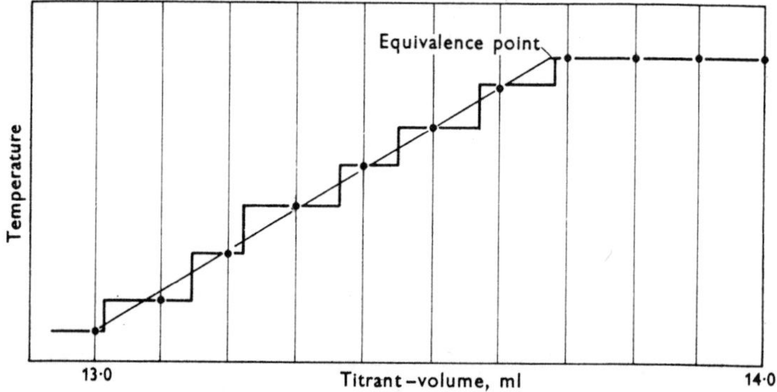

FIG. 2.2 Typical incremental titration curve; from Brown *et al.* (1969).
Vertical lines—chart time–indication lines Horizontal lines—temperature lines.

2.1.2 *Continuous addition*

The continuous addition of the titrant, made possible by the introduction of low heat-capacity temperature detectors, marks the modern phase of thermometric titrimetry. The earliest example of continuous addition was that of Linde, Rogers and Hume (1953), who used a constant-flow burette so that the rate of addition was proportional to the recorder chart speed.

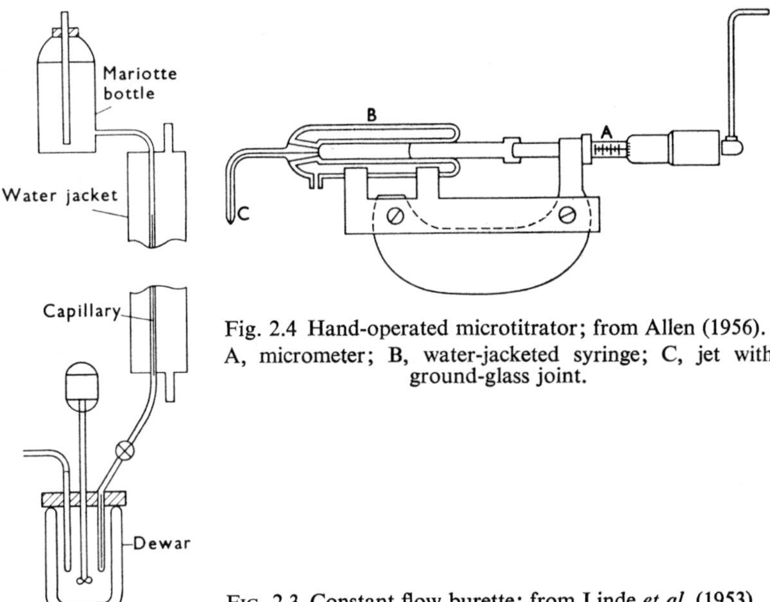

Fig. 2.4 Hand-operated microtitrator; from Allen (1956). A, micrometer; B, water-jacketed syringe; C, jet with ground-glass joint.

FIG. 2.3 Constant flow burette; from Linde *et al.* (1953).

The apparatus shown in Fig. 2.3 had a Mariotte bottle which provided a constant head of titrant. The barrel of the burette contained a length of capillary tubing and both were temperature-controlled by means of a water-jacket. The titrant was therefore delivered at a constant temperature, and because the viscosity of the titrant was constant the flow from the capillary was constant. They found that with the water-jacket at 25·2 °C, 3·105 g of 1·032 M HCl were delivered each minute with a relative standard deviation of $\pm 0.18\%$. In a method using a catalysed end-point procedure, in which the temperature of the titrant was not important, a simple self-filling 10-ml burette was used, with a hypodermic needle butt-jointed to the burette jet to give an approximately constant slow titration rate (S.T.P.T.C., 1967).

Linde and co-workers criticized their own apparatus in that it was not convenient to change titrants and suggested that the motor-driven syringe or piston burette devised by Lingane (1948) should be used, and this titration system has been adopted by the majority of workers since then.

(*i*) SYRINGE OR PISTON BURETTE

The syringe or piston burette, driven by hand, can be used for the incremental addition described above. A typical example is the microtitrator described by Allen (1956) and shown in Fig. 2.4. The titrator consisted essentially of a water-jacketed precision bore tube containing a Teflon piston moved by a hand-turned micrometer.

It is obviously an advantage to use commercial syringes, which can be

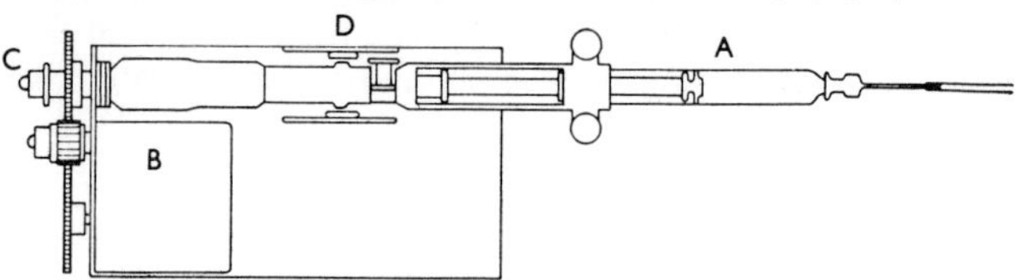

FIG. 2.5 Thermometric titrator; from Vaughan (1970).
A, Agla syringe; B, slow-speed motor; C, slipping clutch for rewind; D, wheeled runner and guide.

obtained in various sizes, in place of the specially constructed one used by Allen. The apparatus used by Jordan (1963a, b) had a syringe clamped to a base, and a screw driven by an electric motor through a gear train was used to push the piston. A similar system employing a Sangamo slow-speed motor to drive an Agla syringe, used by the author, is shown in Fig. 2.5. For reasons which are discussed later it is preferable to use a synchronous motor to drive the screw.

The force required to push the piston can be quite large so it is essential that the syringe is clamped securely and that the base on which it and the screw are mounted is sufficiently solid, otherwise errors are introduced. For the same reason it is essential that the pushing force is in line with the syringe and that, if a dual system is used, the force is balanced evenly between the two syringes (*cf.* Tyson *et al.*, 1961).

The force required to push the piston can also be quite small and in cases where the syringe is used in the vertical position or where the syringe in a horizontal position is much higher than the titration vessel, giving a static head of titrant, the piston should be fastened to the screw to prevent a 'pre-titration' (see Fig. 2.15) or (in the case of the horizontal syringe) the assembly should be at the same level as the titration vessel.

The tubing conveying the titrant from the syringe or burette to the titration vessel can easily be thermostatted by forming it into a coil which is immersed in a constant-temperature

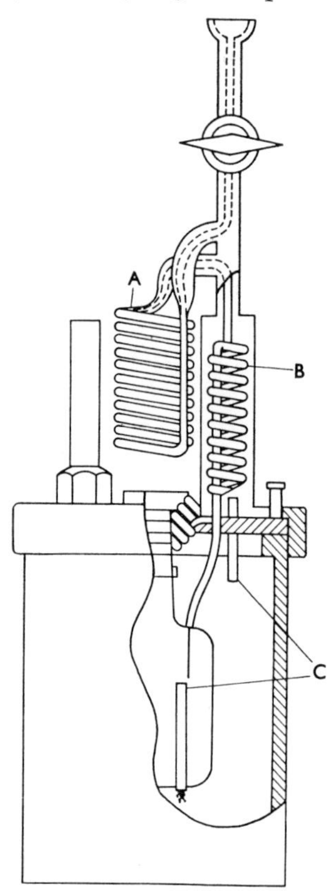

FIG. 2.6 Isothermal jacket titration calorimeter; from Danielsson *et al.* (1964).
A, 2 ml spiral; B, 0·56 ml spiral immersed in mercury; C, 5-junction thermocouple.

bath (see Fig. 2.13). Danielsson and co-workers (1964) passed the titrant through two thin-walled glass spirals as shown in Fig. 2.6: the first, A, had a 2·0 ml capacity; the second, B, had a 0·56 ml capacity and was immersed in mercury that acted as a heat sink to level the periodicity effects of the thermostat which controlled the water-bath in which both spiral A and the mercury were immersed.

Another heat exchanger for the titrant consisted of a thin-walled spiral coiled around a copper coil and enclosed in a eutectic lead–tin mixture. The heat exchanger weighed about 1 kg and served also as a heat buffer to level out fluctuations in the bath temperature (Danielsson *et al.*, 1965).

The titrant can also be thermostatted by placing the whole of the syringe in the constant temperature environment used for the titration vessel, as described by Barthel and Schmahl (1965); see also Fig. 2.4.

In Jordan's (1963a, b) apparatus a 3-way tap is incorporated in the titrant line and connected to a titrant reservoir. By appropriate use of the tap and screw, the syringe can be refilled. Priestley (1963a) modified the syringe and incorporated mercury-weighted non-return valves in the titrant-supply and titration lines so that the syringe operation could be automated.

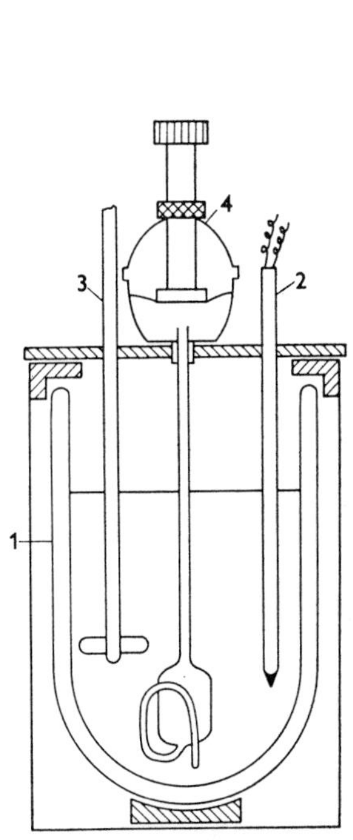

FIG. 2.7 Immersion pipette for single analysis; from Sajó and Sipos (1967b).

1, Dewar; 2, thermistor; 3, stirrer; 4, immersion pipette.

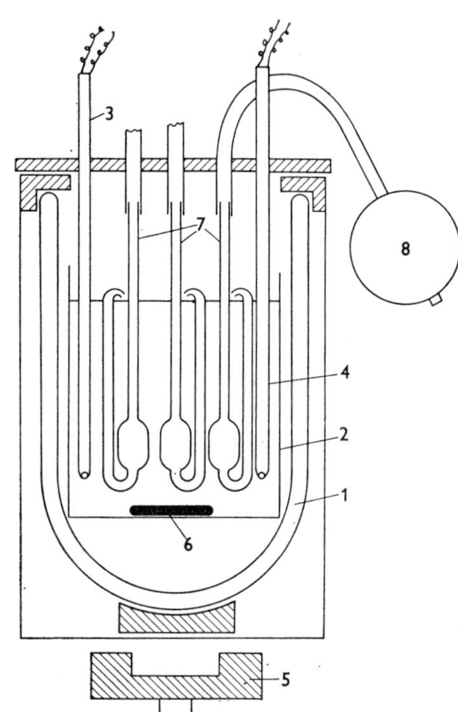

FIG. 2.8 Immersion pipettes for serial analysis; from Sajó and Sipos (1966).

1, Dewar; 2, plastic beaker; 3, thermistor; 4, heater; 5, magnetic stirrer; 6, coated iron follower; 7, immersion pipettes; 8, blowing bulb.

(*ii*) PUMP ADDITION

In continuous-flow enthalpimetry it is essential that the titrant, sample and possibly other reactants are injected into the reaction vessel continuously and as regularly as possible (see Section 1.7). Snyder (1968) used metering pumps, Merényi (1966) inserted capillary jets into the pump lines to give an even flow and Štráfelda and Kroftová (1968a) used either a dosing pump or a triple pump normally employed for amino-acid analyses. Peristaltic pumps used by Priestley and co-workers (1965) and by Taubinger (1969a) are advantageous in that two or more pumps can be driven by a single motor and thus the proportional flow from each pump is not affected by change of motor speed, though care must be taken to ensure a pulseless flow.

2.1.3 *Total addition*

In direct-injection enthalpy the titrant must be added quickly, and most workers have achieved this by the use of pipettes. Figures 2.7 and 2.8

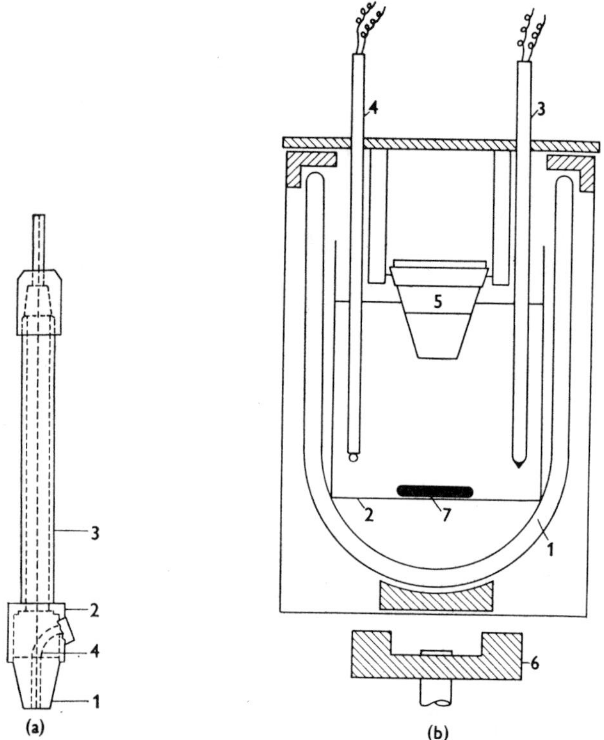

(a)   (b)

FIG. 2.9 Injection apparatus for corrosive materials.

(a) Immersion pipette; from Sajó and Sipos (1968b). 1, platinum crucible; 2, plastic case; 3, 4 plastic tube.

(b) Tipping crucible; from Sajó and Sipos (1966). 1, Dewar flask; 2, plastic beaker; 3, thermistor; 4, heater; 5, platinum crucible; 6, magnetic stirrer; 7, coated iron follower.

show the types of immersion pipettes used by Sajó and Sipos for single and serial analysis.

For corrosive materials Sajó and Sipos used either a platinum/plastic immersion pipette or a tipping platinum crucible as shown in Fig. 2.9. C. E. Johansson (1970) used a glass tube spiral immersed in the titrand with its end finishing as a jet 3 mm above the liquid surface. The spiral held 0·83 ml of titrant and was filled by means of a burette and emptied by an air stream.

2.1.4 *Internal generation*

Vajgand and his co-workers (1970) have developed a method of great potential, using an apparatus in which the titrant is generated cou-lometrically within the titrand (see Fig. 2.10). They applied it to non-aqueous systems in which the end-point is determined catalytically, but this technique can obviously be applied to other systems.

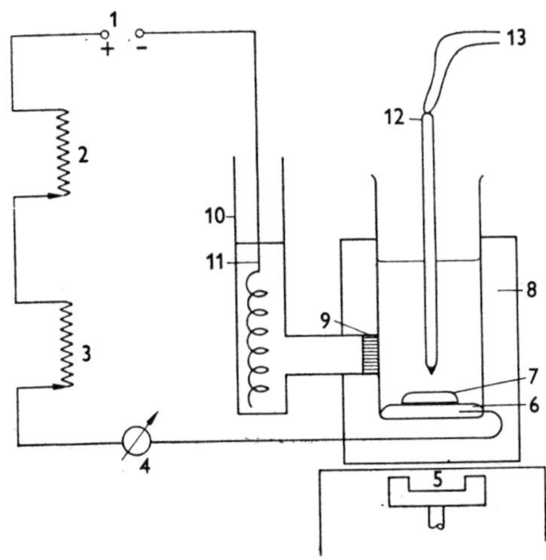

FIG. 2.10 Coulometric—thermometric titration apparatus; from Vajgand *et al.* (1970). 1, stabilized d.c.; 2, 10 kΩ; 3, 100 kΩ; 4, milliammeter, 0.01 mA divisions; 5, magnetic stirrer; 6, mercury anode; 7, coated magnet follower; 8, thermal insulant; 9, sintered glass disc, porosity 4; 10, cathode vessel; 11, platinum cathode; 12, thermistor; 13, to Wheatstone bridge and recorder.

2.1.5 *Addition of the titrant and its precision*

In order to obtain a smooth addition of the titrant in continuous titration methods it is necessary to immerse the burette tip in the titrand. The effect of this is to improve the precision of the addition because errors associated with drop formation are avoided, but at the same time it pro-duces a disadvantage in that titrant–titrand interdiffusion can take place before the titration commences. This diffusion can be minimized by ending the burette in a fine capillary. Danielsson and co-workers (1965) used a 100 × 0·5 mm capillary and the author has found a 25 × 0·4 mm capillary to be satisfactory. In another means of avoiding the diffusion in a very precise titration, the titrant was brought to a particular point along the length of the titrant-entry glass capillary and the volume of capillary not filled was deducted from the final volume (Danielsson *et al.*, 1964). Holmes and Williams (1967b) sealed the end of the capillary with a smear of Apiezon grease before inserting it in the titrand.

The syringe burette has a better precision of addition of the titrant ($\pm 0.10\%$, Tyson *et al.*, 1961, and Raffa *et al.*, 1968; $\pm 0.02\%$, Danielsson *et al.*, 1964) than the continuous-flow burette ($\pm 0.17\%$, Linde *et al.*, 1953) or the incremental addition by burette ($\pm 0.2\%$, Schlyter, 1959). The syringe burette will therefore deliver a smaller volume of more concentrated titrant without loss of precision in the quantity of reactant added. It has been calibrated by weighing 1-bromonaphthalene (which is dense and relatively non-volatile) expelled from it during a specific time (Holmes and Williams, 1967b).

The rate of titration in thermometric titrimetry is not important (see Section 1.4.1) and a wide range of rates has been published and the average of these is about 0.01 ml/s with a range of 0.001–0.3 ml/s.

In an enthalpimetric titration, if the sample solution is injected, the precision of the injection is important, whereas it is less so if the reactant is injected.

2.2 The titration calorimeter

The heat changes of the reaction being studied or utilized in the titration take place in the calorimeter and its design is therefore important. The calorimeter may consist of the following six parts: (i) the titration vessel, (ii) the means of entry for the titrant(s), (iii) the means of distribution of the titrant(s), (iv) the temperature-sensing device, (v) the calibration heater, and (vi) the cooling device. Extraneous heat effects may arise from some of these parts: (i) heat loss or gain from the titration vessel, (ii) mechanical heat gain in the distribution of the titrant(s), and (iii) Joule heat from the temperature sensor. Each of the six parts of the calorimeter is described in the following section together with the methods of dealing with the extraneous heat effects.

2.2.1 *Constant-temperature environment calorimeters*

The simplest constant-temperature environment apparatus, the Dewar flask, was used as the titration vessel in the early titration methods and is still used in recent work by a great number of workers both in thermometric titrimetry (see Fig. 2.3) and in direct-injection enthalpimetry (see Section 2.2.1(*v*)). Although it is satisfactory for many analytical titrations, it has certain limitations, especially when precise values of thermodynamic constants have to be determined or difficult titrations are carried out.

(*i*) FOR PRECISE WORK

The principle disadvantage of the Dewar flask is the length of time needed for thermal equilibrium to be established after a heat change has taken place within it, i.e., it has a poor equilibration time. Sunner and Wadsö (1959) studied the equilibration time of different types of constant-temperature environment calorimeters that they had constructed, by determining the 'heat transfer constant', $\mathbf{K}$, defined by the expression

$$\mathbf{K} = \frac{(dT/dt)_f - (dT/dt)_i}{\bar{T}_i - \bar{T}_f} \qquad (52)$$

To obtain values of $\mathbf{K}$, using Eq. 52, the calorimeter was filled with water and allowed to attain a steady rate of temperature change, $(dT/dt)_i$, with an average temperature, $\bar{T}_i$, over the period. A quantity of heat of about 50 calories was then generated electrically in the water, and after a time, t_f, the temperature change, $(dT/dt)_f$, was determined, having an average temperature $\bar{T}_f$ over the period. The results obtained are given in Fig. 2.11 and show the inferiority of the Dewar flask compared with a properly designed calorimeter.

Sunner and Wadsö concluded from their results that any construction which creates an indeterminate boundary between the calorimeter and its surroundings leads to a slow equilibration. Similar effects are found (*i*) if

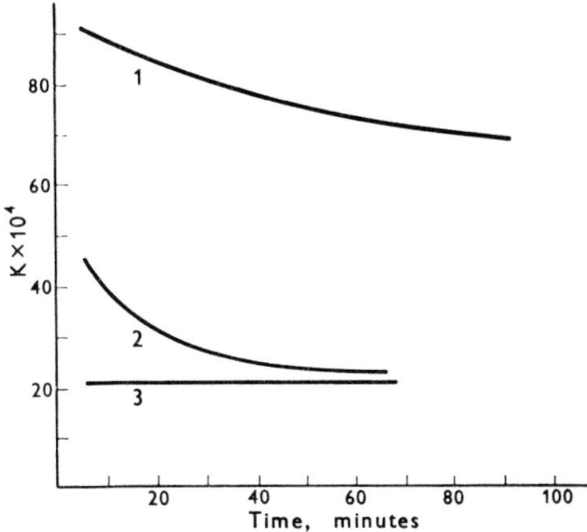

FIG. 2.11 Heat transfer constant, $\mathbf{K}$, *vs.* time, for various calorimeters; from Sunner and Wadsö (1959).
1, ordinary Dewar flask with heavy rubber stopper; 2, steel vessel in Dewar; 3, brass vessel in outer brass can, similar in design to that in Fig. 2.12.

a stopper in good physical contact with both the calorimeter and the surroundings contains insulating materials to slow down the heat transfer, and, (*ii*) if a radiation shield has high heat-capacity compared with that of the calorimeter.

The calorimeters studied in this work were those in which ampoules were broken, but Gerding and co-workers (1963) modified the best calorimeter of Sunner and Wadsö for use in thermometric titrimetry, and this is shown in Fig. 2.12. This calorimeter has a satisfactory equilibration time of three minutes, but between experiments, on cooling, it equilibrated only slowly, taking as long as one hour. In a series of calibration experiments in which new quantities of titrand were used with different quantities of titrant producing a temperature rise of about 0·03 degC, the relative standard deviation of the heat equivalent was in the range 0·10–0·15%.

A precise titration calorimeter incorporating a Dewar flask with a 'plexiglass' taper-joint stopper used by Schlyter (1959) was criticized by Johansson (1965) because the 'plexiglass' expansion was such that it

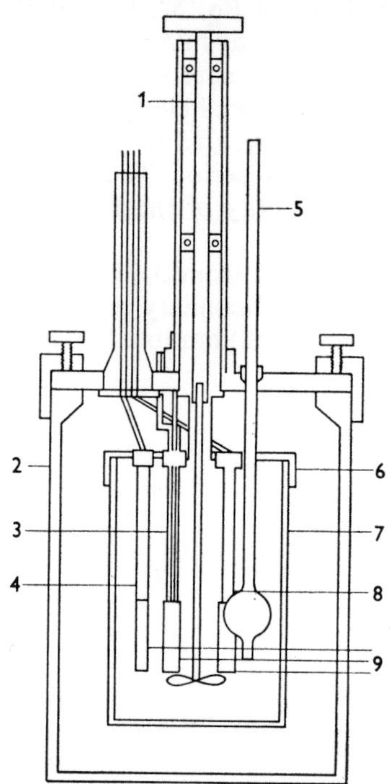

FIG. 2.12 Precise titration calorimeter; from Gerding et al. (1963).
1, stirrer supported by ball races; 2, nickel-plated brass can; 3, cold-gas cooler; 4, thermistor; 5, pipette (see Fig. 2.1); 6, gold-plated brass cover; 7, glass beaker sealed to cover; 8, heater; 9, gold-plated thimbles.

broke the Dewar. He recommended that an inner beaker should be used as the reaction vessel and be made of glass, polythene or gold. The gold beaker was expected to give a low thermal inertia, but at the 1–10 mM levels investigated no significant difference was observed. The titration vessel, in which the lid, made of plexiglass, is held against the silicone-greased ground surface of the 700-ml Dewar by means of springs, is shown in Fig. 2.13. The 300-ml inner beaker is supported by the lid and the whole apparatus is held in the water-bath by means of the stirrer housing. The equilibration time was found to be less than two minutes. A similar equilibration time of two minutes was obtained with a glass titration vessel with a silver lid (Danielsson et al., 1965).

Christensen, Izatt and Hansen (1965), investigating the Dewar as a titration vessel, constructed one with a thin inner wall (0·5 mm) and in the shape of a round-bottomed flask in order to reduce both the heat-leak path and the quantity of glass forming the boundary between the vessel and its surroundings. The vessel, which is shown in Fig. 2.14(a), had a ground-glass socket in which the insert, shown in Fig. 2.14(b), was held. The insert was constructed from a ground-glass cone, and when fitted into the vessel the extension to the cone came within 10 mm of the maximum liquid

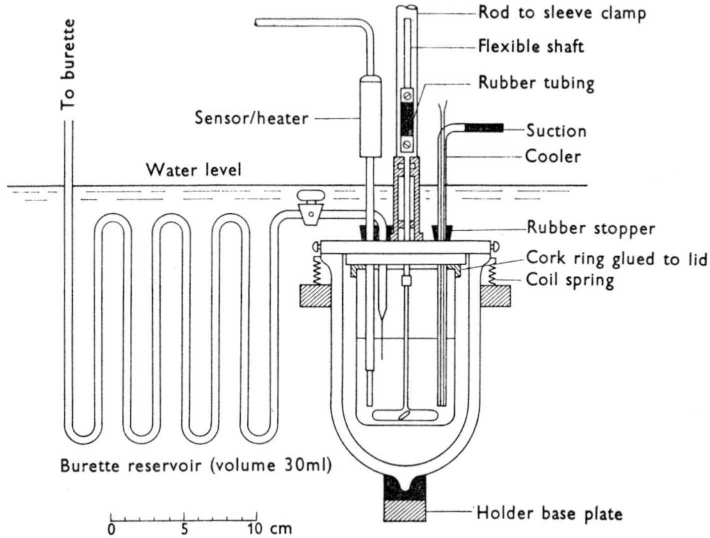

FIG. 2.13 Calorimeter assembly; from Johansson (1965).

level. Polythene discs were pressed into each end of the cone and cemented into place. The discs were drilled to carry the stirrer, titrant delivery tube, thermistor and heater. The inner glass surface of the cone was coated with

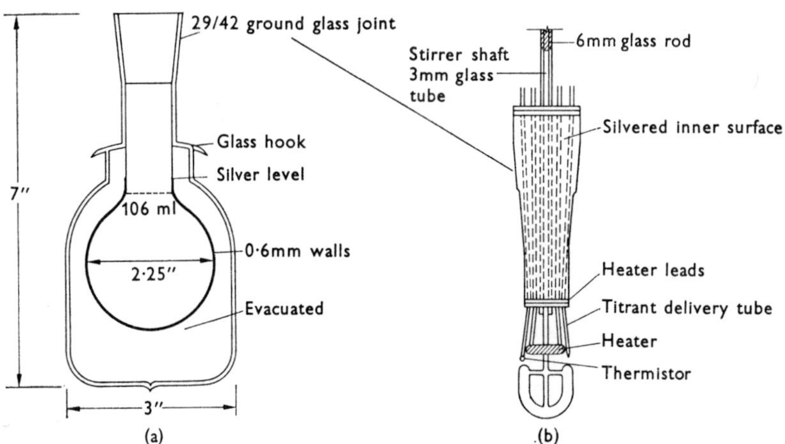

FIG. 2.14 Precise titration calorimeter; from Christensen *et al.* (1965).
(a) Titration vessel; (b) Inset.

several layers of silvering so that good thermal contact was made between the insert and its surroundings.

The resultant calorimeter has an extremely short equilibration time of 1–3 seconds and is thus ideally suitable for continuous titration and also for the determination of heats of reaction as low as 4 cal which can be determined with it to within $\pm$ 0·1% (see also Hansen, 1966).

The insert was modified by Raffa, Stern and Malpas (1968) who used solid Teflon in place of the glass cone. They also used a Teflon stirrer but gave no indication as to whether these modifications improved the performance of the calorimeter or otherwise.

The calorimeters above were immersed in thermostatted water-baths and the temperature variations of these and their method of control are summarized in Table 2.1 in which details of the room thermostatting are also given.

**Table 2.1 Thermostat bath and room conditions
used with precise apparatus**

Temperature variation $\pm$ degC		Control of bath	Reference
Bath	Room		
0·002	—	—	Sunner and Wadsö (1959)
0·004	0·2	mercury contact thermometer	Schlyter (1959)
0·005	—	—	Gerding et al. (1963)
0·001	0·5	mercury contact thermometer	Johansson (1965)
0·001	—	—	Danielsson et al. (1965)
0·003 (o)	—	mercury contact thermometer	Christensen et al. (1965)
0·003 (i)	—	Hallikain thermo-trol 1053A	Christensen et al. (1965)
0·05	0·5	—	Raffa et al. (1968)

(o) outer bath, (i) inner bath

(ii) FOR HIGH-TEMPERATURE WORK

Jordan and his co-workers (1960), investigating titrations in molten salt solution at temperatures of approximately 160 °C, used the apparatus shown in Fig. 2.15 which consisted of an insulated hot-air chamber heated by two electric resistance heaters, the smaller of which was controlled by a thermistor thermoregulator. The chamber contained a turntable on which were mounted four Dewar flasks which could each be raised to bring the titrand into contact with the fixed titration assembly that included an argon gas-purge line.

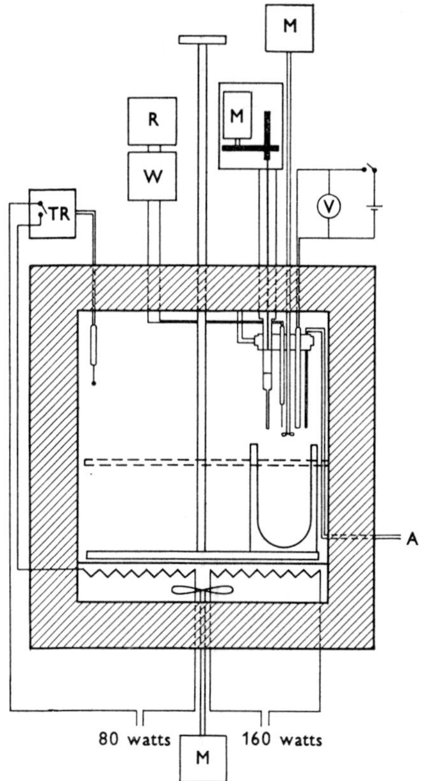

FIG. 2.15 Cross-section of high-temperature apparatus; from Jordan *et al.* (1960a) A, argon inlet; M, motors; TR, thermistor thermoregulator; W, Wheatstone bridge; R, recorder; V, heater circuit.

(*iii*) FOR GAS ANALYSIS

Three types of titration vessel were used by Zambonin and Jordan (1969) in their investigations into gas-injection analysis. They were a 25-ml platinum crucible or a 5-ml borosilicate beaker inserted into a 'Styrofoam' (expanded polystyrene) block or a 250-ml Dewar flask with 3 mm neck. All the vessels had thin walls, necessary for low heat-capacities and rapid thermal equilibration.

The vessels were closed with a silicone bung carrying the gas syringe, calibration heater and thermistor, were stirred with a magnetic stirrer and were used either full or half-full of inert-gas purged titrant for the determination of thermodynamic constants or about a third-full for other analytical titrations (see also Section 2.2.3).

(*iv*) FOR CATALYTIC AND RAPID METHODS

In titrations using catalytic end-points where the degree of adiabaticity is less important simpler titration vessels have been used. Vaughan and

Swithenbank (1965, 1967) used a simple glass tube surrounded by a large one to act as a draught screen as shown in Fig. 2.16 and Vajgand and co-workers (1970) surrounded the coulometric cell with insulating material (see Fig. 2.10). In extremely rapid titrations, where thermal inertia effects

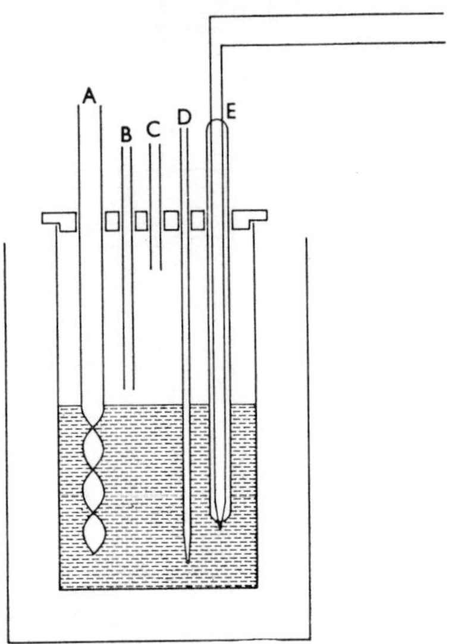

FIG. 2.16 Titration apparatus for catalytic end-point titrations; from Vaughan and Swithenbank (1965).
A, glass stirrer; B, acetone inlet; C, dry air inlet; D, titrant tubing; E, thermistor.

come into play, all that is required is an uninsulated polyolefin-beaker, which is used as a titration vessel, standing on a thermal insulant as shown in Fig. 2.22.

(v) FOR DIRECT-INJECTION METHODS

As in the rapid titration methods above, the direct-injection methods do not require elaborate calorimeters, and most workers have used Dewar flasks; for example those shown in Figs. 2.7, 2.8, and 2.9(b). Other workers have used glass beakers insulated with foamed plastic (Snelson *et al.*, 1967; Beezer and Slawinski, 1971).

2.2.2 *Isothermal calorimeters*

The isothermal titration vessel developed by Christensen, Johnston and Izatt (1968a) as a means of overcoming effects of gain or loss of heat is shown in Fig. 2.17. The important parts of the vessel are the Peltier cooler, connected thermally to the copper cooling cup surrounding the base of the inner titration flask, and the variable input heater. The rate of removal of heat from the flask was set to be greater than the rate of production of heat caused by the addition of the titrant, and the variable heater then kept the temperature constant. The heating process was incremental, the pulse rate being proportional to the difference between the temperature of the titrand and the set temperature of 25 °C in the range from 5×10^{-5} to 2×10^{-4} degC, and thus the temperature of the titrand was controlled to within $\pm 2 \times 10^{-4}$ degC. The number of heat pulses was added and recorded as a function of time.

Because no corrections were necessary for heat exchange between the calorimeter and its environment the results obtained were accurate to

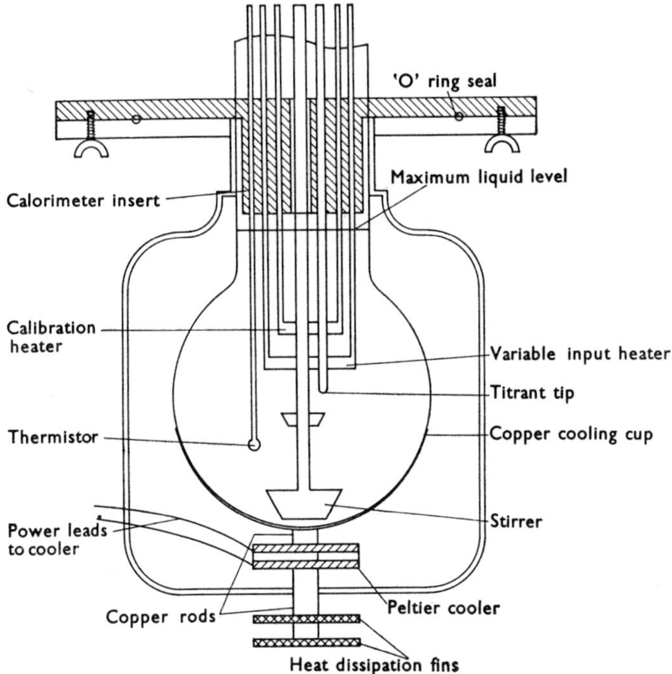

FIG. 2.17 Isothermal titration vessel; from Christensen *et al.* (1968).

within ± 0.02 cal, a better accuracy than that obtained with other types of calorimeters in which corrections have to be made that are often approximate.

2.2.3 *Continuous-flow calorimeters*

In continuous-flow enthalpimetry both the titrant and the reactant have to be brought together continuously under low heat-loss conditions, and because they are both pumped in, the energy of their flows can be used to mix them.

The first apparatus of this type was developed by Niclause in 1946 and reported by Guérin (1952). The apparatus used to determine trace components in gases continuously is shown in Fig. 2.18. The gas to be analysed enters at A and is kept at constant pressure by means of the Dreschel bottle. The flow, measured by the manometer, enters a loop at its centre C where it entrains the reagent and passes along the upper arm of the loop D. The gas-reagent mixture is separated at E and the reagent is returned via a constant head device to the lower arm of the loop F. A thermobattery is used to determine the temperature rise, the cold junctions being placed along the lower loop and the hot junctions along the upper loop.

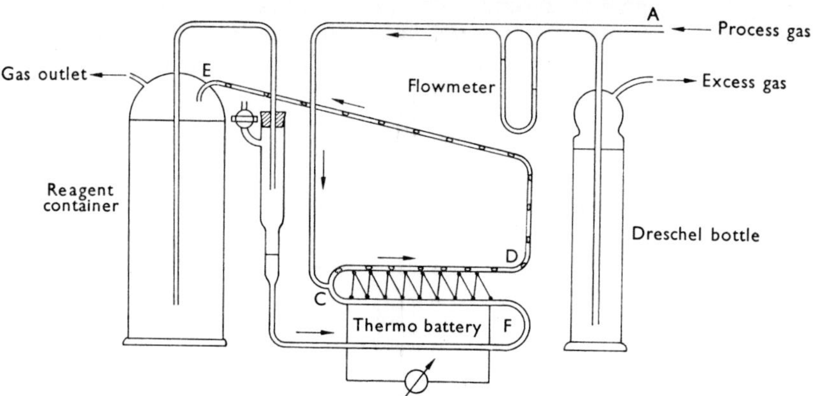

FIG. 2.18 Continuous flow enthalpimetric gas analyser; from Guérin (1952).

Continuous-flow enthalpimetric apparatus was first reported by Lewis (1957) and later by Crespin (1964) but the extensive studies by Priestley, Sebborn and Selman (1965) marked the beginning of the modern phase of continuous-flow enthalpimetry. Their reaction vessel is shown in Fig. 2.19 and consisted of a Perspex cylinder, A, used because of its low heat

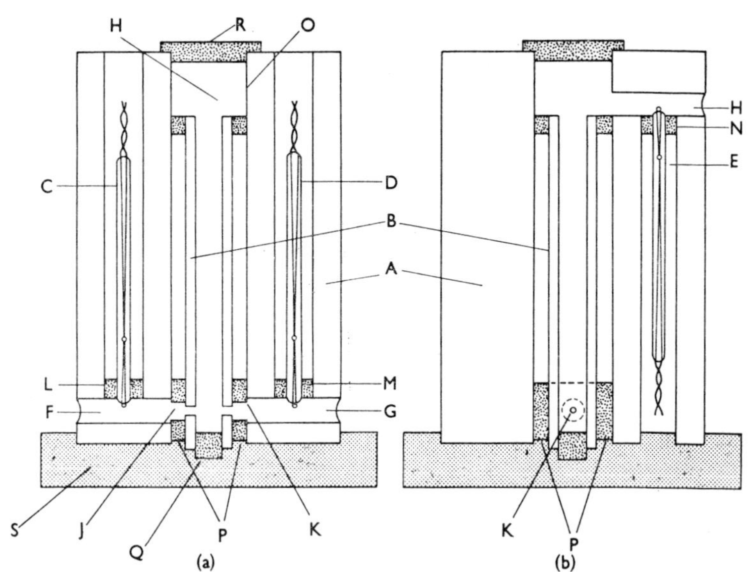

FIG. 2.19 Continuous-flow titration vessel.
(a) front elevation, (b) side elevation; from Priestley, Sebborn and Selman (1965).
A, Perspex cylinder; B, reaction tube; C, D, inlet thermistors; E, outlet thermistor; F, G, inlet tubes; H, outer tube; J, K, opposing mixing jets; L, M, N, O, P, rubber seals; Q, R, rubber plugs; S, polythene plugs.

capacity and thermal conductivity. Two 0·016-in. holes drilled into the reaction tube provided two opposing mixing jets. Other possible types of continuous titration vessels were described in these authors' patent (1968).

Because of the incompatible nature of the latex that was being titrated, Taubinger (1969a) used a concentric entry tube for the sample and reactant, placed at the bottom of a 75 × 20 mm brass tube that contained three equally spaced push-fit baffles. The baffles contained three 1-mm holes drilled 5 mm from the edge and were kept apart by spacers. The tube was isolated from the thermostat bath by placing it in a larger brass tube. A similar baffled reaction chamber was described in a Czechoslovak patent (Štráfelda, 1968).

Priestley and his co-workers (1965) suggested that the rapid mixing apparatus designed by Ruby (1955) should be of advantage because it

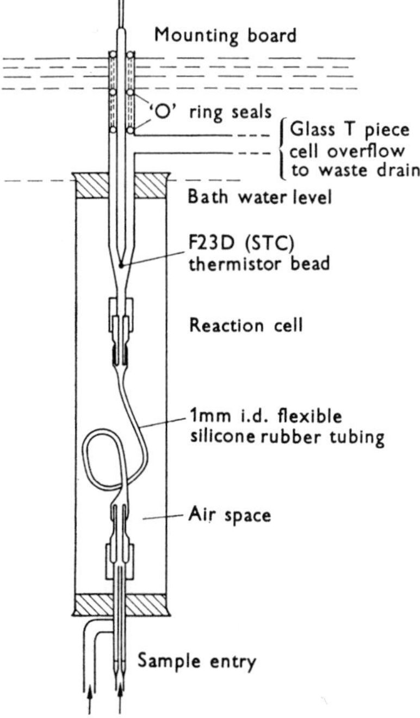

FIG. 2.20 Continuous-flow titration vessel; from McLean and Penketh (1968).

consisted of four tangential mixing jets, two for each solution, which not only gave efficient mixing but were less vulnerable to blocking since the jets could be made larger without loss of efficiency in mixing.

The stirred apparatus used by Snyder (1968) and shown in Fig. 2.21 was constructed of Teflon because of its low heat conduction and inertness towards the organic materials being analysed. It can be seen that no provision was made for measuring the temperature of the reactant, though this could be easily arranged. McLean and Penketh (1968) rejected a stirred apparatus because too much heat was generated by the stirrer, and developed a small heat-capacity, small-volume calorimeter that gave good mixing. Their titration vessel is shown in Fig. 2.20, and it can be seen that the main mixing chamber consisted of a length of silicone tubing in a glass tube maintained at a constant temperature by a water-bath.

A different type of apparatus was used by Štráfelda and Kroftová (1968a) though it has certain similarities to the gas-analysis apparatus described above. Their apparatus, shown in Fig. 2.22, consisted of an aluminium block round which the sample and reagent coils were wound. Cold-junctions of a thermobattery were placed in the tube of one flow from the aluminium block while the tube of the other flow joined the first at a jet beyond the junctions (12, Fig. 2.22). The hot-junctions of the

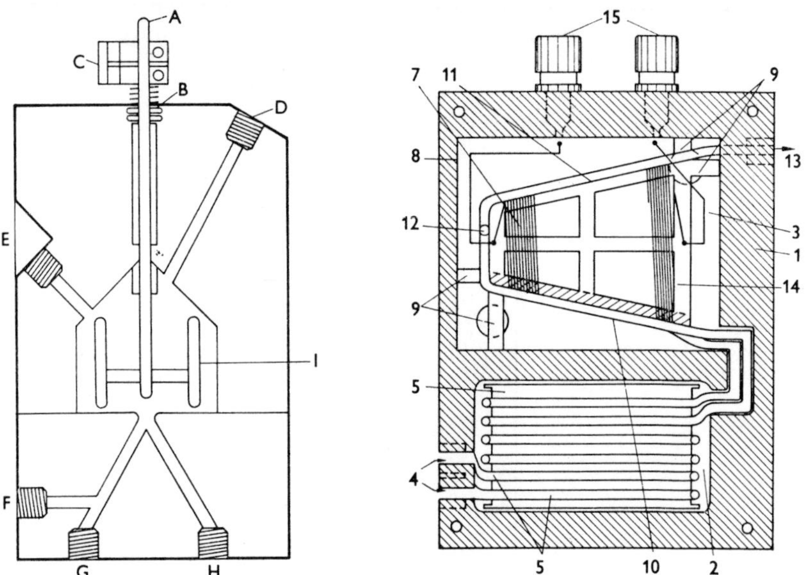

Fig. 2.21 Continuous-flow titration vessel; from Snyder (1968).
A, motor shaft; B, bearing; C, motor coupling; D, outlet; E, reaction thermistor; F, sample thermistor; G, sample inlet; H, reagent inlet, I, 4-baffle stirrer.

Fig. 2.22 Continuous-flow titration vessel; from Štráfelda and Kroftová (1968a).
1, Epoxy resin with 15% powdered aluminium; 2, heat exchanger; 3, thermobattery space; 4, solution inlet; 5, aluminium block; 6, silicone rubber tubings with thin walls; 7, thermobattery; 8, metal casing; 9, spacings; 10, cold junctions of thermobattery; 11, warm junctions of thermobattery; 12, injection opening at junction of tubings; 13, outlet of reaction mixture; 14, supporting frame of thermobattery; 15, thermobattery terminals.

thermobattery were placed on a tube beyond the jet, the tube rising so that any bubbles formed on mixing were carried out of the apparatus. A completely different system was described in a Russian patent (Dobkin et al., 1961). In this the sample flow was kept constant and the titrant flow varied so as to maintain a fixed temperature in the reaction chamber. The rate of titrant feed was then proportional to the concentration of the reactant in the sample flow.

2.3 The stirrer

The stirrer has the combined function of mixing the titrant with the titrand and distributing the heat generated or absorbed by the reaction throughout the reaction mixture. To do this requires a certain amount of energy, which Jordan (1968) estimated could produce 0·05 cal/s when 10 ml of titrand are stirred continuously; the temperature rise of 50 ml of acetic acid was stated by Keily and Hume (1964) to be 8×10^{-5} degC/s. Because it is essential to keep this source of extraneous heat constant so that its effect can be estimated or allowed for by extrapolation of a titration curve, constant-speed motors have been used to drive the stirrer. For rotary-paddle stirrers or magnetic-follower stirrers, motors having rotation speeds of 1–10 rev/s have been used. Bark (1969) recommended the magnetic stirrer (i.e., rotating magnet and coated follower as stirrer) because it had less tendency to produce a vortex than small fast stirrers or interfered less with the fragile thermistors etc. in the calorimeter than did the slower larger stirrers. The magnetic stirrer, however, cannot be used if the calorimeter is immersed well away from the bottom of a constant-temperature bath, and the rotating magnet can be a source of spurious effects in associated electronic recording systems. A variety of types of stirrer have been used, including single and multiple paddle stirrers (see Figs. 2.14(b) and 2.17) and spiral stirrers (see Fig. 2.16). Stirring by means of a gas stream has been used (Charles, 1954) but is not to be recommended because of the evaporative effect. It has been recommended (Raffa *et al.*, 1968) that the titrant entry point and the temperature sensor are placed near each other so that when the titrand is stirred the titrant entering has to make an almost complete circuit before it reaches the temperature sensor. The author has also found this to be the best direction for stirring.

Another method of mixing is by means of the vibratory stirrer. This was used by Priestley and his co-workers (1963) who found that rotary stirrers were ineffective in high-speed titrations. They used a $1\frac{1}{2} \times \frac{1}{8}$ in. flat plate that was a close fit in the titration vessel and which was drilled obliquely with six $\frac{1}{4}$-in. holes around the centre, and outside these with $\frac{1}{8}$-in. holes. A hole through which the thermistor was pushed was drilled in the centre of the plate, as shown in Fig. 2.23. The thermistor was connected to the vibrator which consisted of a unit similar to that in a loudspeaker and was energized at 50 Hz. In an assembled thermometric titrator in which this agitator was used (Priestley, 1963) the vibrator inside the titrator was connected to the thermistor outside by means of a pivoted arm.

2.4 Temperature measurement

The Beckmann thermometer was used for temperature measurement in the early titration method, but because of the size of the bulb it cannot be used in titrand volumes of less than about 45 ml. The bulb size also gives the instrument an appreciable heat capacity, and consequently the titrant can only be added incrementally, to allow time for the thermometer to equilibrate, with the result that the early titration procedures take nearly an hour to complete.

In 1941 R. H. Müller replaced the Beckmann thermometer with the smaller heat-capacity multijunction thermocouple and used this in a comparative method in which heat was generated electrically at a constant rate, in an identical calorimeter, so as to match that produced in the titration calorimeter. The heat generated was timed and the lapsed time

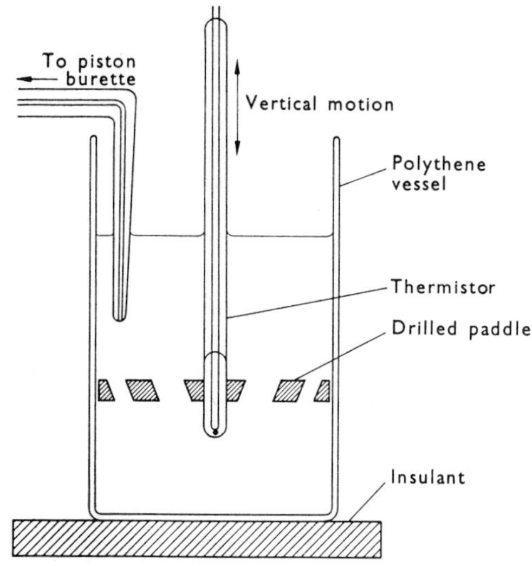

FIG. 2.23　Rapid titration calorimeter; from Priestley *et al.* (1963).

was therefore proportional to the heat evolved in the reaction as a result of the incremental addition of the titrant. The multijunction thermocouple has been used extensively in calorimetry, but only a few workers have used it in thermometric titrimetry (e.g., D. C. Müller, 1957). Nancollas and Hardy (1967) wanted to measure heat changes in solutions of very low concentration in order to facilitate the extrapolation of the thermodynamic data obtained to infinite dilution. They rejected both the thermocouple which they had used previously and also the resistance thermometer in favour of the thermistor with its negligible heat capacity, power dissipation and response time and which, when introduced in conjunction with continuous addition of the titrant in 1953 by Linde, Rogers and Hume, marked one of the most important advances in the development of this technique.

2.4.1　*The thermistor*

The thermistor is a temperature-sensitive resistor with a large negative temperature-coefficient of resistance. The sensor component is a semiconductor composed of ferrites containing traces of other metal oxides, e.g., those of cobalt, manganese, nickel and vanadium, held together with binders (for further details see J. A. Becker *et al.*, 1947; Barron and Rautio, 1968).

The resistance, R_{t_1}, of a thermistor at a temperature t_1 is related to the resistance, R_{t_2}, at another temperature t_2, by the following equation in which B is a constant operating over all but the extremes of the temperature range of the thermistor.

$$R_{t_1} = R_{t_2} \exp B(1/t_1 - 1/t_2) \tag{53}$$

A plot of the logarithm of the resistance against the temperature will therefore give a straight line with negative slope. Values of the resistance of a typical thermistor at various temperatures are: $-50\ °C$—50 kΩ, $0\ °C$—4 kΩ, 20 °C—2 kΩ, 50 °C—700 Ω, 100 °C—200 Ω.

The temperature coefficient of resistance of a thermistor is $-0{\cdot}04$ ohm ohm^{-1} degC^{-1} compared with a value of $+0{\cdot}0038$ for platinum, which is a typical one for a metal and shows the superior sensitivity of the thermistor compared to that of a resistance thermometer. Jordan (1963a, b) stated that a 2-kΩ thermistor (at 25 °C) when wired into a Wheatstone bridge circuit similar to that in Fig. 2.25 would produce an out-of-balance potential of about 40 mV/degC, and that this was equivalent to the response of a 2000-junction thermocouple.

The construction of a typical thermistor used for temperature and flow measurement is shown in Fig. 2.24. The sensor consists of a very small bead composed of the semiconducting material and is usually about

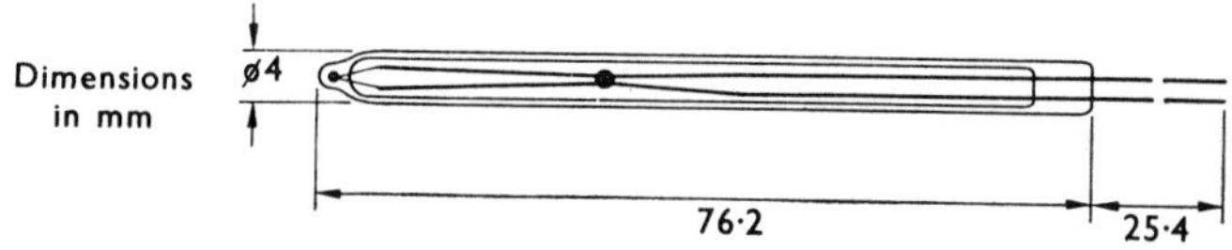

FIG. 2.24 Construction of a typical thermistor for temperature measurement. (By courtesy of Standard Telephones and Cables Ltd.)

0·5 mm in diameter and is connected to platinum leads which in turn are connected to nickel–iron leads. The sensor is sealed into the solid end of the supporting glass tube and covered with a very thin glass envelope to protect it against abrasion or attack. Because of the size of the sensor and its thin envelope the heat capacity is very small and the response time to any temperature change in its environment was quoted by Jordan (1963a) to be less than one second, and it is this property that enabled Priestley and his co-workers (1963) to carry out their very rapid 10–20 second titrations.

The thermistor exhibits slight resistance variations when the pressure and velocity characteristics of the surrounding medium change. The latter is important in thermometric titrimetry and it is preferable to have a constant speed of stirring or of flow in the calorimeter. Müller and Stolten (1953) also found that the exciting potential should be left impressed on the thermistor if a steady state was to be maintained, otherwise several hours were required for the attainment of a new steady state.

Jordan and Alleman (1957) reported that some thermistors gave titration curves with spurious curved portions after the end-point as seen at C–D in Fig. 3.4, and Schlyter (1959) rejected the thermistor in favour of the resistance thermometer because the thermistor resistance at 25 °C changed with time and also altered unpredictably in 'jumps' as was reported by Beck

(1956). Pitts and Priestley (1962) recommended that 'aged' thermistors should be used in order to obtain stable resistance characteristics. They immersed the thermistor in hot and cold water alternately until the resistance at a particular temperature remained constant. Priestley (1965) also recommended 'aged' thermistors for use in continuous-flow enthalpimetry because of their stability.

Assuming therefore that these resistance variations are kept to a minimum and the potential applied to the thermistor is constant, the change in current flowing will be proportional to the change in temperature of the thermistor. The passage of large currents through the thermistor, for example to operate external indicators, will produce self-heating (the maximum is approximately 100 mW) and to avoid this it is necessary to incorporate the thermistor in one arm of a Wheatstone bridge.

To avoid radiation heating of the thermistor from outside the apparatus Gerding, Léden and Sunner (1963) enclosed the thermistor in a tight-fitting thin gold thimble. Malinger and his co-workers (1969) evaluated Czechoslovak thermistors for their possible use in thermometric analysis.

2.4.2 *The Wheatstone bridge*

A number of variations of the basic Wheatstone bridge circuit that are suitable for use with the thermistor have been published. A typical circuit used by Jordan and Alleman (1957) is shown in Fig. 2.25 and consists of (*a*) the voltage supply to the bridge, (*b*) the Wheatstone bridge, and an arbitrary third part (*c*), the recorder zero adjustment circuit.

The variable resistor in (*a*) regulates the sensitivity of the bridge, that in

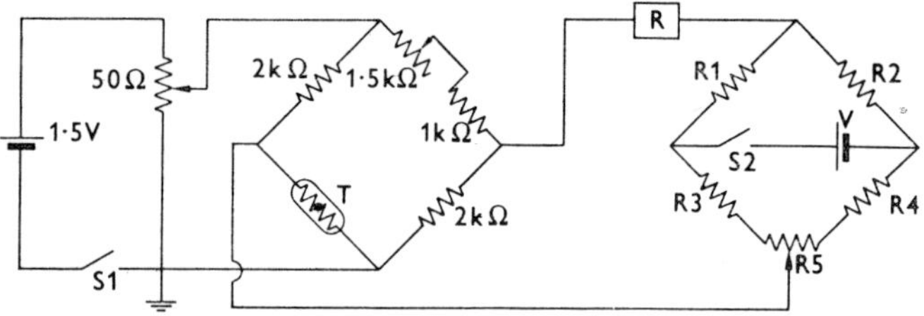

FIG. 2.25 Circuit for use with thermistor and recorder; from Jordan and Alleman (1957). T, thermistor, 2 kΩ at 20 °C; V, low voltage source; R 1–5, resistors chosen to give suitable zero adjustment to the recorder, R; S1, S2, switches.

(*b*) the balance of the bridge at a particular temperature and that in (*c*) the position of the recorder zero. Jordan and Alleman claimed that with this circuit a temperature difference of 0·02 degC could be estimated to within $\pm 1\%$. This temperature rise would be produced by the titration of a 0·002 M solution having a heat of reaction of ± 10 kcal mole^{-1} (see Table 1.1). The circuit was claimed to be linear over a temperature change of ± 1 degC.

This linearity is probably due to the non-linearity of the bridge compensating somewhat for the non-linearity of the thermistor. Pitts and Priestley (1962) constructed a constant-sensitivity bridge in which the bridge-potential resistance was incorporated in the thermistor arm of the bridge as shown in Fig. 2.26.

The total value of the variable resistance, R, was made equal to the change of resistance of the thermistor over the working range and the value of the two variable resistances, Ra, was calculated and set according

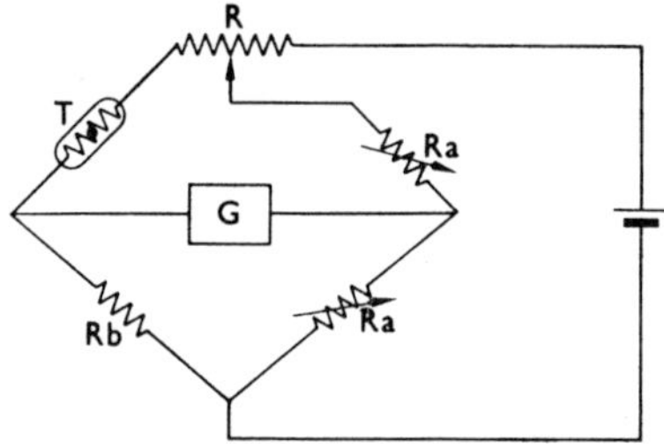

FIG. 2.26 Constant-sensitivity bridge circuit; from Pitts and Priestley (1962).

to the required sensitivity. The sensitivity was checked with a Beckmann thermometer over the range 14·8–28·2 °C and found to vary between 2·45 and 2·47 μA degC^{-1}.

Everson (1967) used a different approach to obtain constant sensitivity with high temperature-change titrations in non-aqueous solution. He proposed (*i*) that the thermistor current should be held constant by making two adjacent arms of the bridge very high in resistance compared to the thermistor and the remaining arm of the bridge and (*ii*) that the change of thermistor resistance should be linearized by shunting it with a resistor whose value is given by the following equation proposed by Nordon and Bainbridge (1962).

$$r_s = r_T (B - 2T)/(B + 2T)$$

where r_s is the shunt resistance, r_T is the thermistor resistance at its mid operation temperature, T (Kelvin), and B is the thermistor constant (see Eq. 53).

Everson showed that by choosing suitable values for the input voltage and high-resistance arms the bridge output per degC could be made exactly equal to the sensitivity of the recorder, thus giving direct temperature readout.

Nancollas and Hardy (1967) described an 800-Hz Wien-bridge transistor oscillator transformer which was coupled to the thermistor bridge. The out-of-balance voltage from the bridge was fed to a high-gain transistor amplifier through a phase-sensitive synchronous detector and an impedance converter. The apparatus was capable of measuring temperature differences of the order of 1×10^{-5} degC and could detect heat changes as low as 3×10^{-3} cal.

In multiple-component analysis by injection enthalpimetry Sajó and Sipos (1966) used a bridge, shown in Fig. 2.27, which incorporated resistors, R_1 and R_2, for use in place of the thermistor for calibration purposes, using the bridge voltage potentiometer, P_2. Switches K_1, K_2 and K_3 were used to obtain full-scale deflection of the galvanometer,

G, when temperature changes of 0·2, 0·5 and 1·0 degC were expected, and the galvanometer shunt resistor, R_3, to obtain an expansion of 10 on the galvanometer scale. A series of preset resistors were engaged by the switches $K_{SiO_2\%}$ etc. These preset resistors were such that, with an appropriate temperature switch (K1–3), the galvanometer gave a full-scale

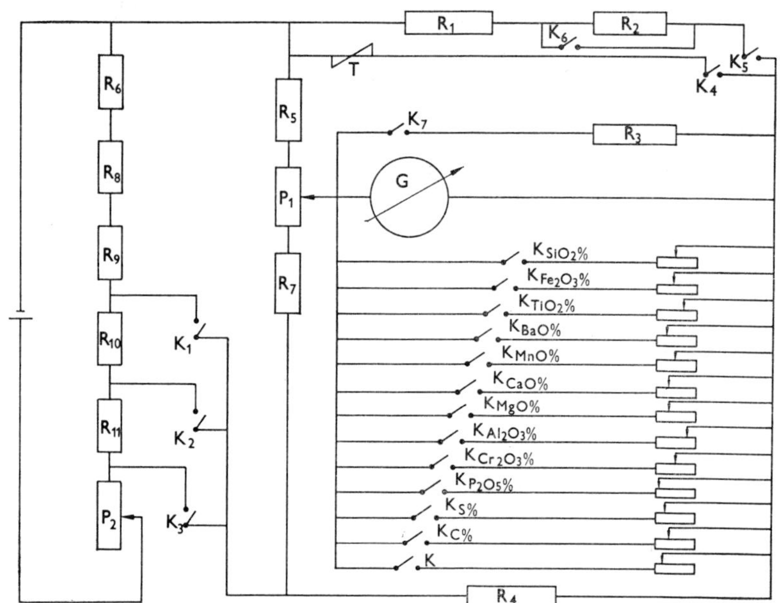

FIG. 2.27 Wheatstone bridge for multiple injection enthalpimetry; from Sajó and Sipos (1966).

T, thermistor; G, galvanometer; R, resistors; P, potentiometers; K, switches.

deflection for 10% of the particular component, e.g., SiO_2 or, if switch K_7 were engaged, 100%.

In a later paper Sajó and Sipos (1968d) used rotary selection switches for the component resistors and also for the temperature range resistors which, in this case, were used as galvanometer shunts.

In continuous-flow enthalpimetry it is necessary to compare one temperature sensor with the mean of the outputs of two other sensors (see Section 2.12). Priestley (1965) described a bridge circuit incorporating linearized transistors (see above) which not only performed this function but also incorporated a calibrated potentiometer which backed off the out-of-balance current of the bridge, and the setting of which was linearly related to the required temperature difference. The circuit shown in Fig. 2.28 has been patented (Priestley et al., 1968).

It is important that the two thermistors W and Z should be a matched pair (or matched according to the method of Godin, 1962), and for the bridge to balance, thermistor P should have a higher resistance than that of W or Z at an equivalent temperature. The resistance of the preset resistor M is varied by trial and error until the out-of-balance potential remains constant when all three thermistors are immersed together in water at 15, 20 and 25 °C. The circuit is set up by turning D to its lower limit, passing water and reactant through the calorimeter, and bringing the galvanometer to zero by using the control B. The water is then replaced by a test solution of known concentration and the value of D changed to an appropriate setting, e.g., a 'Digidial' value of 500 for a 0·500 M solution, and the galvanometer again brought to zero, this time by using the control C.

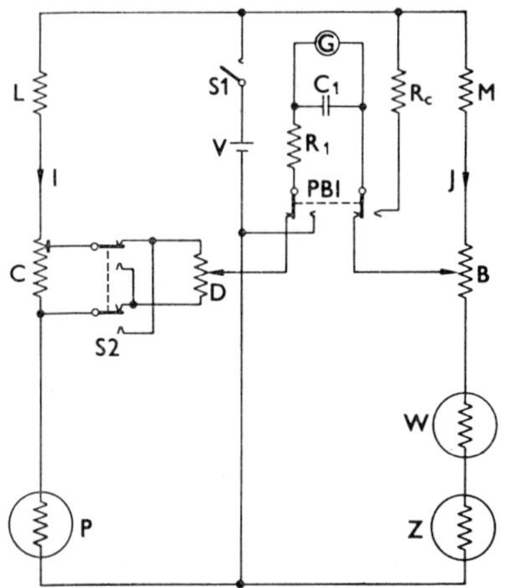

FIG. 2.28 Thermistor bridge circuit for continuous-flow enthalpimetry; from Priestley (1965); Priestley *et al.* (1968).

V, 90 V; L, 100 kΩ; M, 200 kΩ (preset); R_i, 10 kΩ, C_i, 1 μF (damping); R_c, 150 MΩ, for voltage check; S1, on-off switch; S2, reversing switch; PB1, push-button for voltage check; G, galvanometer, 5·5 nA mm^{-1}, 1·4 kΩ; C, sensitivity potentiometer, 100 Ω single turn; B, balance potentiometer, 1 kΩ ten turn; D, calibrated backing-off potentiometer, connected to a 'Digidial', 1 kΩ ten turn; P, W, Z, linearized and aged thermistors.

With an unknown solution passing through the calorimeter the galvanometer is again brought to zero with control D; the reading on the 'Digidial' then gives the required concentration.

Priestley (1963b) automated this procedure by replacing the galvanometer by a servo-amplifier which in turn operated a servo-motor coupled to the potentiometer and associated 'Digidial'. The rotation of the 'Digidial' indicated in a digital manner the angular position of the potentiometer shaft so that each digit represented 0·1 degC. Thus with appropriate adjustment of the circuit it should be possible to obtain a direct reading of content on the 'Digidial'.

2.4.3 *Recording the temperature*

The use of thermometers in the early titration method meant that temperature readings had to be taken after each addition of the titrant and the reading plotted against the appropriate titrant volume to obtain the graph. R. H. Müller (1941) stated that unless results were automatically recorded or evaluated the method was too time-consuming. The combination of continuous addition of the titrant and of electrical temperature-sensing, permitted the continuous recording of the titration. Thus with chart recorders the titration curve is produced continuously and can be interpreted immediately the titration is completed. Provided the chart advancement rate and the titrant addition rate are both constant in a thermometric titration, the length of chart will be proportional to the amount of titrant added. These conditions can be easily achieved with motorized addition of the titrant if the titrant motor and the recorder motors are both synchronous.

For thermometric titrations with rounded end-points Zenchelsky and Segatto (1957b) suggested that the first and second differentials of the temperature-time functions would improve the precision of the end-point. They used a transmitting slidewire coupled to the chart recorder and used resistor–capacitor circuits to effect the differentiation. Perchec and Gilot (1964) used the derivative circuit of a recording titrator, and Bark and Doran (1969) used a similar circuit of a d.c. polarograph, both monitoring the output of a Wheatstone bridge. Because the signal level is low and the rate of voltage change is small, thermal 'noise' caused by the stirrer and electronic noise gave rise to very irregular titration curves, so suitable filtering circuits had to be incorporated. Unfortunately none of these authors made any comparisons of the results obtained for the same titration with and without differentiation.

2.5 Calibration of the calorimeter and its temperature adjustment

In order to calibrate a calorimeter so that enthalpy changes can be calculated from the measured temperature changes it is necessary to generate a known quantity of heat within the calorimeter under the volume conditions of the reaction being studied. This can be carried out by the use of a Joule heater or less satisfactorily by performing a titration under similar calorimeter conditions, using a reaction of known enthalpy.

2.5.1 *The heater*

For results of the highest accuracy it is essential that the heat is generated in the heater at such a rate that it matches the rate of heat evolved in the titration. This of course only applies to exothermic reactions, for in an endothermic reaction the heater is used to supply heat to the system during the course of the reaction so that the overall temperature change at any one time is very small. The heat supplied by the heater under these conditions is equal to the heat absorbed in the reaction.

Schlyter (1959) constructed a heater, using 0·07-mm manganin wire as the heating element, which he enclosed in a glass tube filled with oil to conduct the heat to the outside. Johansson (1965) found that this use of a continuous glass tube gave a persistent heat loss of about 3% and recommended the heater shown in Fig. 2.29 in which the heat is conducted effectively to the outside via a tight gold casing fitted to the heating element. The heating element consisted of a 15-Ω length of silk-covered manganin wire and was aged for four hours at 110 °C before use.

The heater element described by Stern *et al.* (1966) would appear to be ideal for use in thermometric titrimetry. It incorporates a 15-Ω, 0·25-W carbon resistor and a Zener diode power circuit, is compact and has a rapid response.

In order to calculate the amount of energy liberated in the heater (in J degC^{-1} or cal degC^{-1}) it is necessary to measure the current flowing in the heater, the potential drop across the heater and the time that the current flows. To obtain the highest precision in these measurements it is necessary (*i*) to measure the current flowing at intervals during the heating because the resistance of the heater changes with change of temperature, (*ii*) to ensure that the applied potential (usually from an accumulator bank) is constant, by altering a variable resistance in parallel with the source (alternatively constant-voltage variable-setting d.c. supply apparatus is now available), and (*iii*) to use a mechanical or electrical connection so that switching the heater on or off immediately performs the same action on the timer.

In a precise calorimeter, similar to that in Fig. 2.14, which has a short equilibration time, a heating period of 90 minutes has been shown to produce an error of about 0·3%. In less precise apparatus the heating time must therefore be reduced to a period of several minutes to obtain the same level of accuracy. The heating period should preferably be within 10–20% of the end-point time in the experiment for which the calibration is being carried out (Carr, 1972).

A typical circuit for the calibration heater is shown in Fig. 2.30. The ballast resistor is similar in resistance to the calibration heater. Initially, the ballast resistor is connected until the voltage of the battery is constant, then the switch is shifted so that the calibration heater is in circuit and the timer is also switched on. The values of R_1 and R_2 are each approximately 1 kΩ (Schlyter, 1959) or higher to give an appropriate reading on the 'potential' potentiometer, P_p. The resistor, R_s, is a standard for measuring the current, using the 'current' potentiometer, P_s.

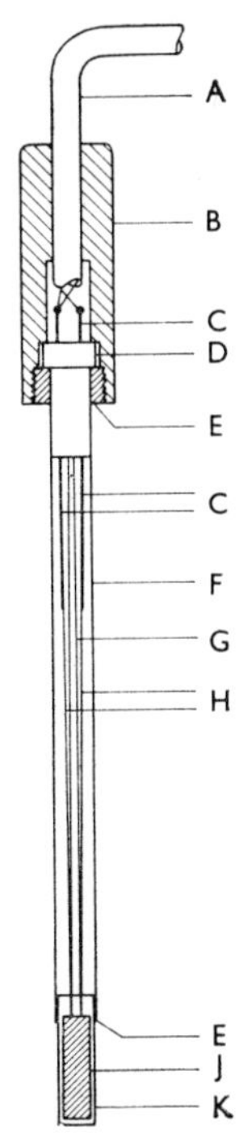

FIG. 2.29 Joule heat calibrator; from Johansson (1965).

A, 4-lead cable; B, threaded plexiglass plug; C, copper tag; D, machined plexiglass inset; E, epoxy resin ('Araldite') glue line; F, outer glass tube; G, inner glass tube; H, copper lead; J, plexiglass cylinder with silk spun manganin wire; K, gold casing.

2.5.2 *Calibration titration*

The most common calibration titration procedure is that in which a strong acid is neutralized with a strong base, for example, Mead (1962) titrated an appropriate volume of 0·02 M HCl with 1 M NaOH. A value of −13·42 kcal/mole (or −56·15 kJ/mole) can be taken (Hale *et al.*, 1963) for the enthalpy of this reaction and used with the observed temperature rise in the calculation.

Wilson and Smith (1969) criticized this procedure because of the disadvantage that the enthalpy of the reaction, though high, varies with the concentration and also because the sodium hydroxide titrant can easily pick up carbon dioxide from the atmosphere. They proposed that the sodium hydroxide should be replaced by tris(hydroxymethyl)aminomethane (THAM or Tris) which is available from the U.S. National

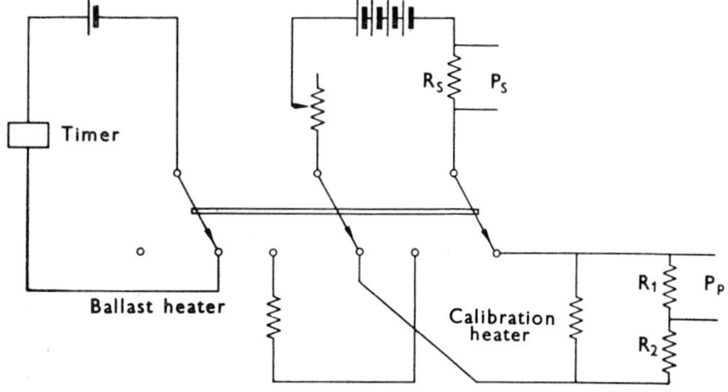

FIG. 2.30 Calibration heater circuit; after Schlyter (1959).

Bureau of Standards as a calorimetric standard. In their titration of 6 g of THAM with 0·1 M HCl no change of ionic strength was observed and they reported the value of the enthalpy of the reaction to be $-11·35 \pm 0·06$ kcal/mole (see Table 4.20). The work of Hansen and Lewis (1971b) showed, however, that the titration of THAM with perchloric acid in aqueous solution was less reliable than a similar titration of sodium hydroxide with perchloric acid and they recommended the latter reaction as a thermometric standard.

Alexander and his co-workers (1969) used the unusual but highly exothermic reaction of iron(II) with cerium(IV) ($-\Delta H = 27·0$ kcal/mole) as a thermometric standard.

2.5.3 *The cooler*

Because no convenient cooling device is available that gives a quantitative knowledge of the amount of heat extracted by it, a cooler can only be used to adjust the temperature of the calorimeter and its contents. If a number of measurements of exothermic reactions are made and perhaps heater calibrations are also carried out in the calorimeter, then a negative slope for the post-reaction period of the titration occurs and precise measurement of the reaction heats becomes impossible and a cooler placed in the calorimeter is then necessary. A cooler can also be used to lower the titrand temperature to that at which the measurement is to be made or to a standard temperature (see Fig. 1.5). Schlyter (1959) used as a cooler a simple concentric tube through which air at 0 °C was drawn (or carbon dioxide at 0 °C; Gerding *et al.* (1963)). Johansson (1965) placed a droplet of water in a similar but gold-plated thimble and drew air over it (see Fig. 2.13). This method was stated to be more rapid than the cold-air system, a large cooling effect being obtained in less than a minute.

In order to cool the titrand in serial analyses involving reactions with a high enthalpy change, Hoffmann and Tornau (1962) used a Dewar flask that could be evacuated and which after the reaction was filled with hydrogen, and was re-evacuated when the temperature of the titrand had dropped. Spink and Spink (1968) used the simple expedient of inserting a cold glass rod into the titrand in order to cool it.

2.6 Extraneous heat effects

One of the most common difficulties in titrations involving heat changes is caused by the need to have a relatively concentrated titrant, which is likely to produce a large heat of dilution. In thermometric titrations this can obscure the end-point (but see Section 1.8.4) and in enthalpimetric titrations can produce a large blank value which, if it opposes the reaction enthalpy change, can be very serious. The heat generated by the stirrer (see Section 2.3) and the Joule heat from the thermistor or thermopile are other unwanted heat effects. Singly these may be small, but combined may be significant especially if the reaction has a low enthalpy change or if it is carried out in a system of low heat-capacity, such as a small volume.

In thermometric titrations it is possible to minimize the effects by the use of differential titrations and it is also possible to prepare special titrants with low or even zero heats of dilution.

2.6.1 *Differential titration*

One solution to the problem of unwanted heat effects in thermometric titration is to use a mathematical transformation of the titration graph from rectangular to oblique co-ordinates to allow for these extraneous heat effects (Popper *et al.*, 1965). Another possibility is to use a differential titration as proposed by Tyson, McCurdy and Brecker (1961).

In this procedure, two identical syringes driven by a single screw thread delivered the titrant to two identical titration vessels one of which contained the sample. The motor speeds for driving the stirrers could be adjusted to balance the stirrer heat effects, and matched thermistors in each titration vessel were connected into opposing arms of a Wheatstone bridge, the complete circuit of which is shown in Fig. 2.31.

The resistances R_1 and R_4 were adjusted to compensate for any difference between the heat capacities of the two vessels, on the basis of the heat generated (by means of a calibration heater) and not on the temperature rise. Thus the only extraneous heat effect is that caused by the slight difference, whilst the reaction is taking place, in the Joule heats of the thermistors when the sample vessel is at a different temperature from that of the blank.

Extremely sharp end-points were obtained even at low concentrations; Keily and Hume (1964), using a similar system, obtained a relative standard deviation of less than $\pm 0.2\%$ in non-aqueous titrations of metal

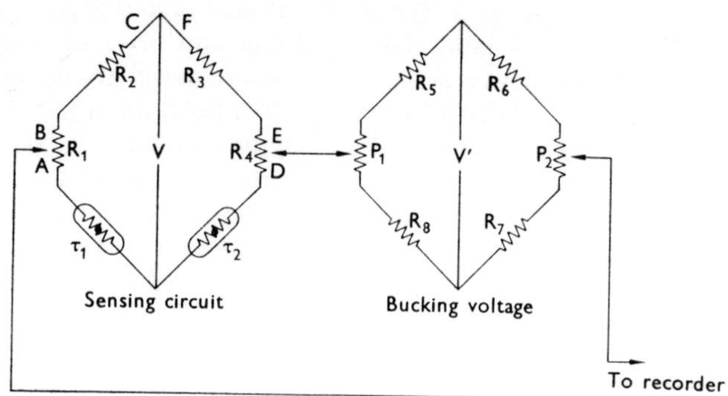

FIG. 2.31 Differential titration circuit; from Tyson *et al.* (1961). P_1, 50 Ω ten-turn potentiometer; P_2, 1 kΩ potentiometer; R_1, 4 kΩ potentiometer; R_2, 16 kΩ resistor; R_3, 18 kΩ resistor; R_4, 2 kΩ potentiometer; R_5 to R_8, 5 kΩ resistors; V, variable voltage from dry cells, 0 to 22, 5 V; V', 6-V dry cell; $\tau_1 = \tau_2 =$ sets of four 51 Al thermistors in parallel. Resistance at 25 °C 25 kΩ.

acetates with perchloric acid. Rondeau, Legrand and Pâris (1966) used a simplified differential titration in which they added titrant only to the vessel containing the sample. The blank vessel containing the reference thermistor was only stirred, and in this way they claimed that they compensated for the heat of stirring and Joule heat from the thermistor.

2.6.2 *Graphical interpretation of the titration curve*

When the temperature change in a thermometric titration has to be measured in order to obtain data for calculating the enthalpy change of the reaction, graphical methods can be used to allow for certain of the

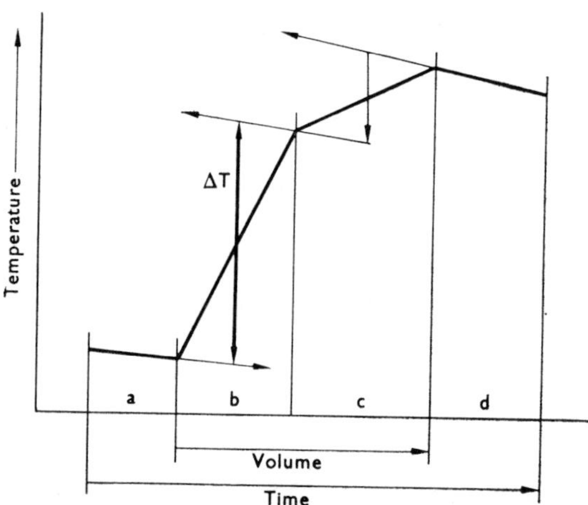

FIG. 2.32 Hypothetical titration curve showing graphical interpretation; after Barthel *et al.* (1968).

(a) base period—no titrant added;
(b) reaction period—titrant added;
(c) excess titrant period—titrant added;
(d) post titrant period—no titrant added.

extraneous heat effects. Figure 2.32 shows a hypothetical titration curve in which the titrand is losing heat to the surroundings and the titrant is warmer than the titrand at the end of the titration.

The figure illustrates the method, which is to extrapolate backwards from the post-titration period (d); to draw a vertical line bisecting the excess titrant period (c) to meet the extrapolated line and extended downwards the same distance to join this lower point to the point at the end of the reaction period (b); to extrapolate forwards from the base period (a); finally to bisect the reaction period (b) with a vertical line to meet the extrapolated and extended lines. The length of this final line is the one required.

The titration curve is also affected by phase changes during the titration, in particular the formation of precipitates, which can have a marked

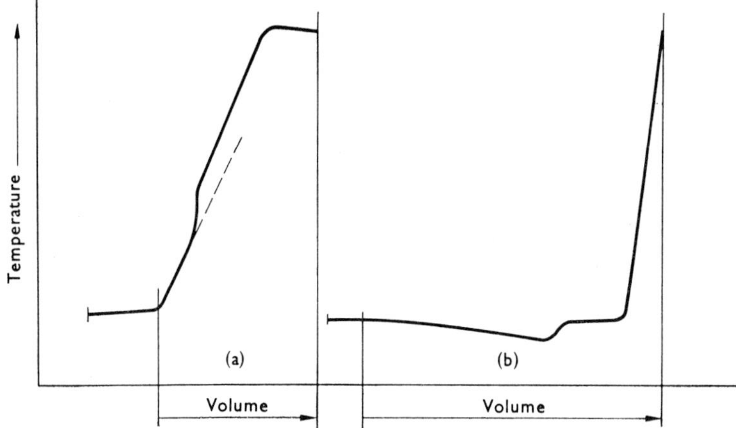

FIG. 2.33 The effect of delayed precipitation during the titration.
(a), aniline with 0·1 M HBr both in acetonitrile; from Forman and Hume (1964).
(b), p-t-butylphenol in acetone with M KOH in propan-2-ol; from Vaughan (1970).

effect as shown in Fig. 2.33. The delayed precipitation of aniline hydro-bromide in the titration of aniline with hydrogen bromide, both being in acetonitrile solution, is shown in Fig. 2.33(a) and that of the potassium salt of p-t-butylphenol in the titration of the latter with M KOH in propan-2-ol, with acetone used as the thermometric indicator, is shown in Fig. 2.33(b).

The titration curves of other phenols which affected the end-point in a similar titration are shown in Fig. 4.3 (p. 200).

2.6.3 *Titrants with low heat of dilution*

In enthalpimetric titrations compensation for extraneous heat effects can be made by a second injection of the reagents. This injection is made when the temperature conditions have become stable after the first injection. The temperature rise from this second injection is then sub-tracted from the first, in which the reaction takes place. The conditions are

not quite the same on the second injection so the method is only approximate. The method also requires more time and to improve this Sajó and
his co-workers have devised titrants with low or zero heats of dilution for
use in serial-component enthalpimetric analysis. For example in the
determination of silica by the injection of 40% hydrofluoric acid (Sajó
and Sipos, 1966) a large heat of dilution was obtained. This is illustrated
in Fig. 2.34 in which it can be seen that linear calibration lines were

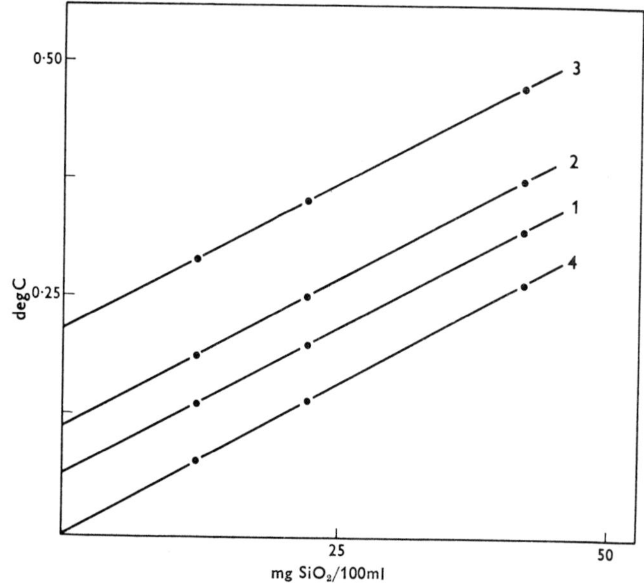

FIG. 2.34 Calibration lines for the determination of silica with 40% HF solutions; 1,
1 ml; 2, 2 ml; 3, 5 ml and 4, 8 ml of 40% HF/urea solution; from Sajó and Sipos (1966).

obtained when the temperature rise was plotted against the silica content.
The lines however had a larger intercept on the temperature change axis
the larger the quantity of titrant added, because of its high heat of dilution.
This meant that the quantity of titrant injected had to be constant, which
is not desirable in an enthalpimetric titration. Sajó and Sipos showed that
if urea were added to this titrant, then because urea solutions have an
endothermic heat of dilution, the resultant mixture could be adjusted to
have a zero heat of dilution. This is shown by the line in Fig. 2.34 that
passes through the origin regardless of the fact that far more of the titrant
was injected.

Table 2.2 gives an outline of the ways, apart from adjusting the dilution
of the titrant, that Sajó and Sipos have used to prepare titrants with low
heats of dilution. Further details of the method are given in the appropriate
sections. This work has been concerned with enthalpimetric titrations but
there is no reason why the principle cannot be applied to titrants for
thermometric titrations.

Table 2.2 Methods of preparation of titrants with low heats of dilution
(By Sajó and Sipos)

Titrand	Titrant	Temperature-adjusting substances		Reference section
Al(III)	HF	Urea	*or* HF	3.10.2
Al(III)	NaCl	$N(C_2H_4OH)_3$	*or* HCl	3.10.3
B(III)	HF	$CO(NH_2)_2$	*or* HF	3.9.1(*i*)
Ba(II)	H_2SO_4	NH_4Cl	*or* H_2SO_4	3.8.2
Ca(II)	$K_2C_2O_4$	KOH	*or* $(COOH)_2$	3.6.2(*ii*)
Cl^-	$AgNO_3$	$AgNO_3$	*or* C_2H_5OH	3.27.1
Cr(III)	$KMnO_4$	NaCl	*or* $KMnO_4$	3.34.2(*ii*)
Cr(VI)	$FeSO_4$	$FeSO_4$	*or* H_2SO_4	3.41.1(*i*)
Cu(II)	KCN	KCN	*or* Glycerol	3.30.6(*ii*)
Fe(II)	$(NH_4)_2S_2O_8$	H_2O	*or* $(NH_4)_2S_2O_8$	3.36.1(*v*)
Mg(II)	Na_2HPO_4	NaOH	*or* HCl	3.5.2
Pb(II)	Na_2S	H_2O	*or* Na_2S	3.17.7
Si(IV)	HF	$CO(NH_2)_2$	*or* HF	3.15.1
Sn(II)	HF	$CO(NH_2)_2$	*or* HF	3.16.2(*ii*)
SO_4^{2-}	$BaCl_2$	$BaCl_2$	*or* HCl	3.24.5(*ii*)

Note. The left-hand alternative temperature-adjusting substance lowers
the heat of dilution.

2.7 Automatic apparatus

The automation of thermometric or enthalpimetric titration apparatus
began with the continuous addition of the titrant from a flow burette and
later with the use of the syringe burette. It continued with the use of
electrical temperature detection and the associated recording of the
temperature–time (titrant) titration curve. Full automation was achieved
by Priestley (1963a) who used a digital timer to measure the time of the
titration, which was proportional to the amount of reagent added (see
Section 2.1.2(*i*)), the rate of addition being constant. A transistor circuit
converted the temperature change (which commenced at a constant rate
on addition of the titrant) into a linear voltage change. This was in turn
converted to give a square-wave voltage, the amplitude of which was
proportional to the rate of temperature change, the period being equal to
the duration of the change. The end-point was detected by a transistor
relay circuit which switched on with the change of voltage at the beginning
of the titration and switched off when the change of voltage ceased. De Leo
and Stern (1965) amplified the out-of-balance potential from the Wheatstone bridge. The potential was then filtered, differentiated and applied to
the control unit of a Sargent–Malmstadt automatic titrator. The latter was
used to amplify and differentiate the signal twice and was then able to

trigger a relay system to stop the titration on the then abrupt signal-change at the end-point.

Continuous-flow enthalpimetry is obviously the ideal means of automatically producing results. McLean and Penketh (1968) have described apparatus that was capable of determinations in the 0·005–0·1 M range and Taubinger (1969a) one capable of monitoring large temperature changes associated with high concentrations.

Guillot (1970) described a completely automatic apparatus based on injection enthalpimetry. The apparatus is shown diagrammatically in Fig. 2.35, and it can be seen that a programmer is used to operate two

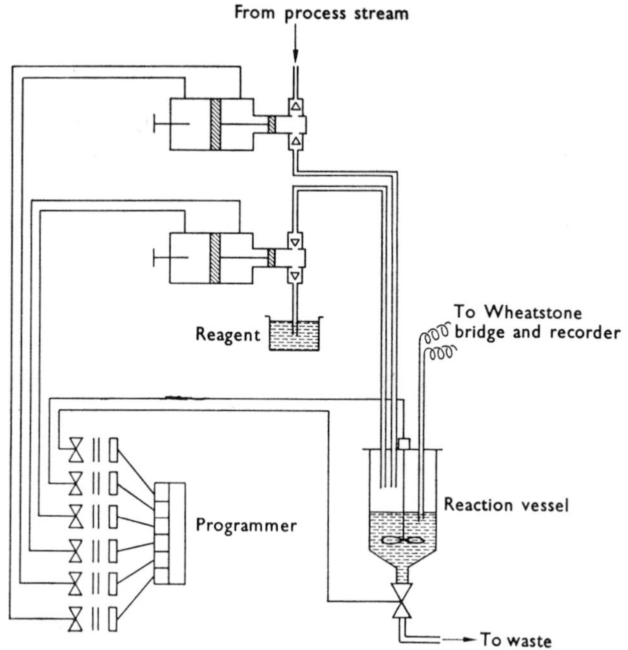

FIG. 2.35 Automatic injection enthalpimetric apparatus; from Guillot (1970).

pneumatic pipettes, the reaction vessel stirrer and emptying valve. The following sequence is used: (i) close vessel valve, (ii) add sample, (iii) start stirrer, (iv) add reactant, (v) record temperature change, (vi) stop stirrer, (vii) open vessel valve. The results can obviously be given in digital form from this type of apparatus, which was used for both discontinuous or repetitive analyses with a precision of about 1%.

2.8 Commercial apparatus

The first commercial thermometric titration apparatus was described in a United States patent by Wasilewski (1964) and assigned to the American Instrument Co. Inc. The latter company market the instrument under the name 'Titra-Thermo-Mat' and it is illustrated in Plate I and shown diagrammatically in Fig. 2.36. Essentially, the instrument consists of (*i*) a synchronous motor driving a syringe burette that is mounted vertically and connected to a digital counter that provides a direct readout of titrant volume (ml); (*ii*) a titrant reservoir from which the burette is automatically refilled and in which is incorporated a thermistor to measure the temperature of the titrant; (*iii*) an 'adiabatic space', which contains the burette, erected on the base of the instrument, an L-shaped stirrer connected to an external motor, the titrand thermistor and a heater element, the whole

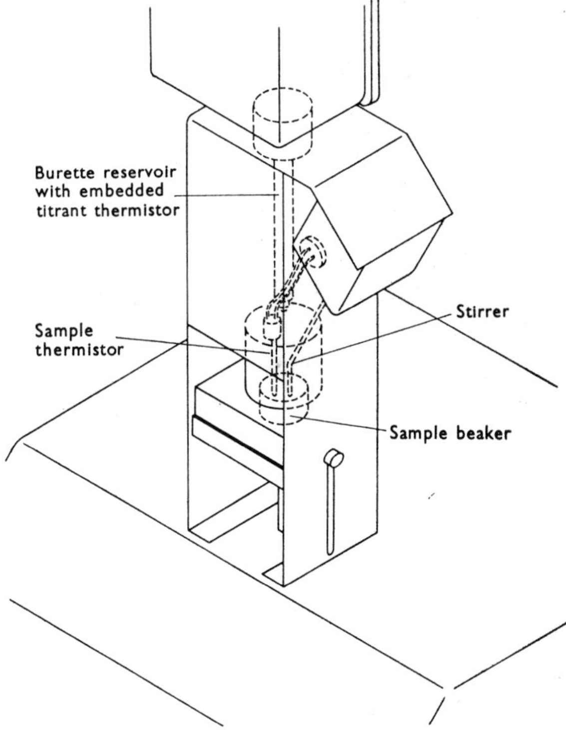

Fig. 2.36 Adiabatic space of the Titra-Thermo-Mat.
(By courtesy of the American Instrument Co. Inc.)

being insulated with expanded polystyrene; (*iv*) a platform on which the beaker (30 or 20 ml capacity) is placed and raised into the 'adiabatic space'. The burette tip, the thermistor, stirrer blade and heater are then correctly positioned in the titrand.

LKB Instruments Ltd. market a thermostatted system in which a number of different types of calorimeters can be inserted, one of which is a titration calorimeter. This calorimeter, illustrated in Plate II, was based on a design by Scandinavian workers (Danielsson *et al.*, 1964) and has a capacity of 100 ml. Up to 5 ml of titrant can be introduced through the coil of Teflon tubing to ensure that it is thermostatted. The calorimeter is calibrated after each reaction and it is stated that a heat change of 20 cal

PLATE I

Aminco Titra-Thermo-Mat

(By courtesy of the American Instrument Co. Inc.)

will have a relative standard deviation of less than 0·02%, i.e., an uncertainty of 0·004 cal.

A constant-temperature environment titration calorimeter is marketed in the U.S. by Tronac Inc.

Two patents describing apparatus for multiple-component analysis by the direct-injection enthalpimetric method have been published (Sajó *et al.*, 1966 and Sajó and Sipos, 1969a) and these have been assigned to the Magyar Optical Works, who market the apparatus under the name Directhermom. This apparatus, illustrated in Plate III, is based on the work of Sajó and Sipos described in Section 2.1.3, in the latter part of Section 2.4.2 and in Section 2.6.2 (see also Sajó, 1968c). The instrument contains 24 variable resistors which can be set to cover the determination of 24 various components over the range 0·1–100%. The apparatus has been used to determine nine components of slag, using a single weighing of sample and taking 3–5 minutes for each component.

A commercial enthalpimetric flow microcalorimeter, capable of measuring the velocity of reactions under the conditions given in Section 1.7.1, is available from LKB Instruments Ltd., as designed by Wadsö and Monk (1968).

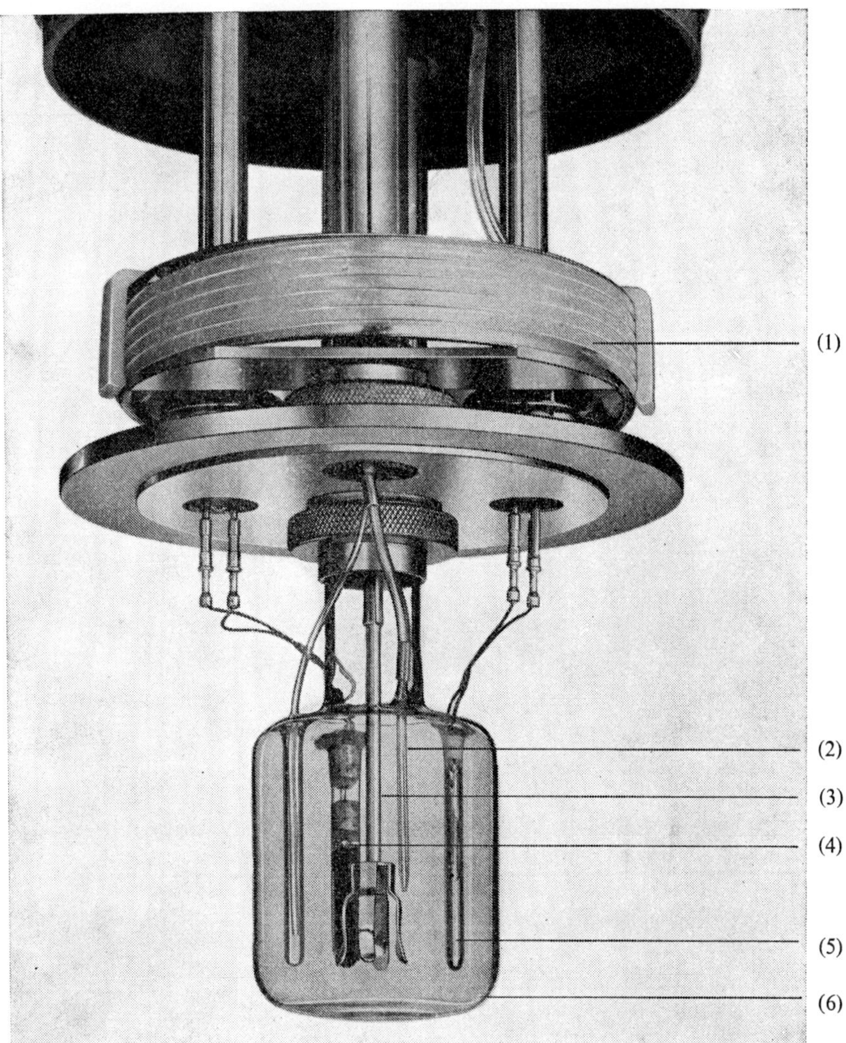

PLATE II

Titration Calorimeter (part of the LKB 8700 Precision Calorimetry System)
Reproduced by permission of LKB Instruments Ltd.

(1) Teflon tube containing thermostatted titrant
(2) Capillary through which titrant is added to reaction vessel
(3) Stirrer to ensure complete mixing of reactants
(4) Heater, for electrical calibration of calorimeter
(5) Thermistor, for measurement of temperature change
(6) Reaction vessel

PLATE III

DIRECTHERMOM multiple component direct injection apparatus
(by courtesy of the Magyar Optical Company)

Chapter 3

Inorganic Reactions

To enable an easy comparison to be made between the reactions of different elements or compounds and to facilitate the easy extraction of information relating to one particular element or compound, this chapter has been arranged according to the Periodic Table, i.e., in the order s^1, s^2, p^1, p^2, p^3, p^4, p^5, d^n, f^n elements, followed by neutralization reactions. The final sections are concerned with metal alkyls, general properties of metal-ion complexes of aminopolycarboxylic acids, and the important catalytic end-point method applicable to determination of traces of anions or cations.

Metal compounds containing organic radicals are covered in this chapter if their reaction and properties are concerned primarily with the metal, otherwise they are discussed in Chapter 4.

3.1 Lithium

3.1.1 *Reaction of lithium with* EDTA

The heat of reaction of lithium ions with EDTA, found by a direct-injection method, was very low (Charles, 1954) and because of this the attempts to determine lithium by titration with M EDTA, using an automatic digital titrator, were unsuccessful (Priestley, 1963a).

Note

The titration of lithium acetate and chloride is discussed in Section 4.1.4(*ii*) and of butyl lithium under Metal Alkyls (Section 3.50); for the titration of lithium nitrate see Section 3.18.3.

3.2 Sodium

The thermometric studies of sodium compounds are almost entirely concerned with the associated anions and these are considered in the appropriate sections. The determination of sodium as NaK_2AlF_6, discussed below, is a good example of how a little-known reaction can be utilized in an enthalpimetric titration.

3.2.1 *Determination of sodium as* NaK_2AlF_6

If sodium ions are added to a solution containing the compound K_3AlF_6, precipitation of NaK_2AlF_6 takes place with the evolution of heat. This unique reaction has been utilized by Sajó (1969a) for the determination of sodium, using a direct-injection enthalpimetric method.

The reagent is prepared by dissolving pure aluminium in hydrochloric acid and adding potassium chloride and hydrofluoric acid. The reagent is saturated with NaK_2AlF_6 by adding a small quantity of sodium chloride. The sample containing sodium ions is injected, from a platinum pipette, into the reagent which is held in a plastic beaker.

When a solution containing 107 mg of Na_2O was analysed in the presence of 100, 200 and 500 mg of potassium chloride and then of 100, 200, 500 and 1000 mg of ammonium chloride, the results were correct to within ± 2 mg of Na_2O. Table 3.1 shows the results obtained on known solutions.

Table 3.1. Determination of sodium by an enthalpimetric method

(From Sajó (1969a))

Na_2O *added*, mg	Na_2O *found*, mg	*Difference*, mg
214·0	217.0	+3·0
107·0	105·0	−2·0
53·5	55·0	+1·5
26·8	28·0	+1·2
13·4	13·0	−0·4
6·7	6·1	−0·6
3·3	3·8	+0·5
1·7	1·3	−0·4
0·8	0·5	−0·3

3.3 Potassium

The thermometric titration studies of potassium compounds are almost entirely concerned with the associated anions and these are considered in the appropriate sections. A number of thermometric precipitation titration procedures for the determination of potassium have been investigated by French workers using the early titration methods (Rondeau *et al.*, 1966). Of possible titrants, perchloric acid and sodium hexanitritocobaltate(III) gave too slow a precipitation, dipicrylamine and picric acid were not soluble enough to provide convenient titrant solutions; calcium picrate in alcoholic solution was ineffective; only fluosilicic acid and sodium tetraphenylborate were effective, as described below.

3.3.1 *Determination of potassium as* K_2SiF_6

Potassium-containing solutions of concentrations down to 0·02 M, mixed with twice their volume of alcohol and saturated with K_2SiF_6, have been titrated with 0·5 or 1·0 M fluosilicic acid or zinc fluosilicate, with an accuracy of between 0·5 and 1%. The ions Ca^{2+}, Ba^{2+}, Sr^{2+}, Al^{3+}, which form insoluble fluosilicates, must be absent and the concentration of sodium ions must not be greater than that of the potassium. The ions Li^+, Ag^+, Cu^{2+}, Zn^{2+}, Mg^{2+}, Mn^{2+} and Ni^{2+} were found to have no effect.

3.3.2 *Determination of potassium as* $K[B(C_6H_5)_4]$

The solubility of sodium tetraphenylborate in water is too low for it to be possible to obtain a concentration sufficient for a thermometric titrant. This problem was solved by adding a known excess of sodium tetraphenylborate to the potassium-containing sample at a pH between 4 and 6. The solution was stirred, allowed to come to thermal equilibrium and then titrated with a concentrated solution of potassium chloride. A rapid analysis with an accuracy of 1% was achieved with solutions of concentrations above 0·01 M. It was shown by Carr (1971a) that with the more sensitive thermistor temperature detection, it was possible to use a direct titration with 0·5 M sodium tetraphenylborate as titrant. Samples with potassium concentrations in the range 12–30 mM were determined with a precision and accuracy of better than 1%, but with concentrations as low as 6 mM the end-points were rounded. The titration was found to be independent of pH in the range 2–13, and it was shown that at pH 12 there was no interference from ammonium ions. Mercury, silver and thallium interfered but not iron(II), zinc or cadmium if the titration were carried out in 0·2 M EDTA at pH 8.

Carr (1971b) also used tetraphenylborate to determine the enthalpy of precipitation of potassium and other univalent ions. The values of the enthalpies found, in kcal/mole, were NH^+_4 8·24; K^+ 10·05; Rb^+ 12·34; Cs^+ 12·67; Tl^+ 17·28 and Ag^+ 19·69. The errors in these determinations ranged from $\pm 0·09$ to $\pm 0·28$ kcal/mole.

3.4 Beryllium

3.4.1 *Determination of beryllium as the* EDTA *chelate*

Priestley (1963a) has determined beryllium nitrate by titration with M EDTA, using an automatic digital titrator. Six titrations had a relative standard deviation of $\pm 1·4\%$. The reaction in this case was endothermic. It is interesting to note that only an indirect indicator titration method has been developed for the determination of beryllium with EDTA (Misumi and Taketatsu, 1959).

3.4.2 *Beryllium systems forming complex ions*

The early titration method has been used to study complex beryllium ions found in solutions. The results are summarized in Table 3.2.

3.5 Magnesium

3.5.1 *Determination of magnesium as the* EDTA *chelate*

The stability constants of magnesium and calcium EDTA chelates are close in value, so in the indicator titration method, if the indicator reacts with

Table 3.2 Summary of studies of complex beryllium ions

Titrant	Titrand	Complex ions formed	Reference
FLUORIDES			
NH_4F	BeF_2	$[BeF_4]^{2-}$	(a)
NH_4F	$(NH_4)_2BeF_4$	$[BeF_5]^{3-}$; $[BeF_6]^{4-}$	(b)
NH_4F	BeF_2	$[BeF_3]$; $[BeF_4]^{2-}$; $[BeF_5]^{3-}$; $[BeF_6]^{4-}$	(b)
CHLORIDES			
NH_4Cl	$BeCl_2$	$[BeCl_3]$; $[BeCl_4]^{2-}$; $[BeCl_5]^{3-}$; $[BeCl_6]^{4-}$	(c)
BASIC SULPHATES			
$NaOH$	$BeSO_4$	$BeSO_4.BeO$	(d)
$BeSO_4$	$NaOH$	$H_n[Be(OH)_nSO_4]$; $(BeOH)_2SO_4$	(d)
PYROPHOSPHATES			
$BeSO_4$	$Na_4P_2O_7$	$[Be(P_2O_7)_2]^{6-}$; $[Be(P_2O_7)]^{2-}$	(e)
$Na_4P_2O_7$	$BeSO_4$	$[Be(P_2O_7)_2]^{6-}$; $[Be(P_2O_7)]^{2-}$; $?[Be_2(P_2O_7)]$	(e)

(a) Chatterji (1958b); (b) Purkayastha (1947);
(c) Novoselova *et al.* (1955); (d) Haldar (1948b); (e) Haldar (1950b)

both ions, e.g. Eriochrome Black T, both ions are titrated together and cannot be distinguished. From this it would appear that the ions could not be distinguished in a thermometric titration, but magnesium, unlike the calcium and many other EDTA chelates, has an endothermic heat of chelation with EDTA, and it is possible therefore to titrate magnesium directly after the titration of, or removal of, calcium if it is present (see Fig. 3.1). The interference of other ions and multiple-ion titrations are discussed in Section 3.49.

Jordan and Alleman (1957) titrated 25 ml of 0·01 M Mg^{2+} and obtained a precision of 0·4% with an error of 0·5%. They quoted a lowest limit of determination of 2 mM with an error of 3%.

Using an automatic apparatus, Priestley and co-workers (1963) titrated 0·25 mmole of Mg^{2+} in 20 ml of solution and obtained a T value of $-0·1$ degC/mmole; they also titrated magnesium sulphate, using a continuous flow apparatus (Priestley *et al.*, 1968).

The results obtained by the use of other automatic titrators are given in Table 3.3 and show that they are equal, if not superior in precision to the indicator titration method, a view expressed by De Leo and Stern.

Wasilewski, Pei and Jordan (1964) used the EDTA titration of magnesium to evaluate the performance characteristics of direct-injection enthalpimetry which they were developing. Injecting 300 μl of titrant into 25 ml of magnesium solution of concentration between 1·95 and 9·92 $\times$

10^{-3} M they obtained a relative standard deviation of $\pm 2\%$ with an error of the mean of 3%. The titrant was 1 M EDTA (Na_4 salt) in all the cases above.

3.5.2 *Determination of magnesium as phosphate*

This well-known precipitation method for magnesium has been utilized in thermometric titrimetry. It requires the same sample preparation as in gravimetry to remove the materials precipitated by ammonia/ammonium

Table 3.3 Precision of tests in the automatic determination of magnesium

Mg^{2+} concentration	Replicates	Deviation, %	References
0·017 M	5	$\pm 1 \cdot 6$	Priestley (1963a)
0·006 M	5	$\pm 1 \cdot 6$	De Leo and Stern (1965)
0·009 M	5	$\pm 1 \cdot 0$	De Leo and Stern (1965)
0·012 M	5	$\pm 0 \cdot 5$	De Leo and Stern (1965)

chloride and also calcium precipitated by, or titrated thermometrically with oxalate, but in spite of this is considerably faster than the classical gravimetric method.

Chatterji (1955) determined magnesium in dolomite, using the early titration method, first titrating the calcium thermometrically with oxalate (see Section 3.6.2) and then adding excess of ammonia and titrating with approximately 0·5 N $NaNH_4HPO_4$ (microcosmic salt). The results were consistently 1% lower than those obtained by the gravimetric method and this was attributed to loss of magnesium in the calcium titration by post-precipitation in the presence of the excess of oxalate titrant.

Recent work with use of the direct-injection method has shown that it is possible to determine rapidly magnesium in slags, cements and clinkers (Sajó and Sipos, 1966, 1968a, 1969b), in ores, rocks and industrial silicates (Sajó, 1969c), and in magnesites and dolomites (Sajó and Sipos, 1968d). The samples were prepared as described under Manganese (Section 3.35), and 5 ml of saturated potassium permanganate, followed by 8 ml of near-saturated potassium oxalate, were added to the sample solution to prevent interference from manganese and calcium; alternatively, the solution after the titration of calcium (Section 3.6.2(*ii*)) was used. The prepared sample was injected with 8 ml of a nearly saturated solution of either diammonium or disodium hydrogen phosphate. The nearly saturated solution (950 ml of saturated solution diluted to 1 litre) was modified, if necessary, to give no heat of dilution when 8 ml were injected into a blank solution; however if the temperature fell, small amounts of M HCl were added to the titrant, and if the temperature rose, either ammonia or 2% sodium hydroxide (as appropriate to the phosphate used) was added until no heat of dilution occurred. A series of 12 replicate determinations of a sample known to contain 41·33% MgO had a range of 41·2–41·7% with an average of 41·43%. The results obtained are shown in Table 3.4.

Table 3.4 Determination of magnesium as MgO in slags, clinkers, cements, magnesites and dolomites

(Comparison of methods; from Sajó and Sipos (1966, 1968a, 1968d))

Sample		MgO, % Complexometric method	Enthalpimetric method
Blast-furnace slags	1	4·1	4·3
	2	3·8	3·7
	3	7·7	7·9
Siemens-Martin slags	1	7·0	7·2
	2	10·6	10·3
	3	9·1	8·9
	4	9·0	9·1
Clinkers	1	3·1	3·1
	2	4·2	4·4
	3	1·3	1·5
	4	2·5	2·5
Cements	1	2·5	2·3
	2	2·6	2·8
Magnesites	1	41·3*	41·4
	2	43·9*	43·7
	3	41·7*	41·3
Dolomite		19·7*	19·9
Burnt dolomite		43·1*	43·5

* Method not stated

It can be seen that the results do not show the consistent negative error found by Chatterji, using the early titration method; they compare well with the complexometric indicator titration method. Details of the precipitation of magnesium oxalate are discussed in Section 3.6.2.

3.5.3. *Formation of basic salts of magnesium*

Dutoit and Grobet (1921) studied the basic salts of magnesium nitrate by using the early titration method. They titrated 20 ml of N $Mg(NO_3)_2$ with 6 M NaOH and obtained a curve with three inflections which were attributed to the formation of $MgNO_3.OH$, $Mg(OH)_2$, and possibly $Na_2[Mg(OH)_4]$.

Note

The titration of magnesium acetate is discussed in Section 4.1.4(*ii*)

3.6 Calcium

3.6.1 *Determination of calcium as the* EDTA *chelate*

As might be expected from the results in the complexometric indicator titration method, all workers titrating calcium thermometrically with 1 M EDTA (Na$_4$ salt) report that in the presence of magnesium, the whole of the calcium is titrated before the chelation of magnesium starts (see Fig. 3.1). The reason for this is discussed under Magnesium (Section 3.5.1).

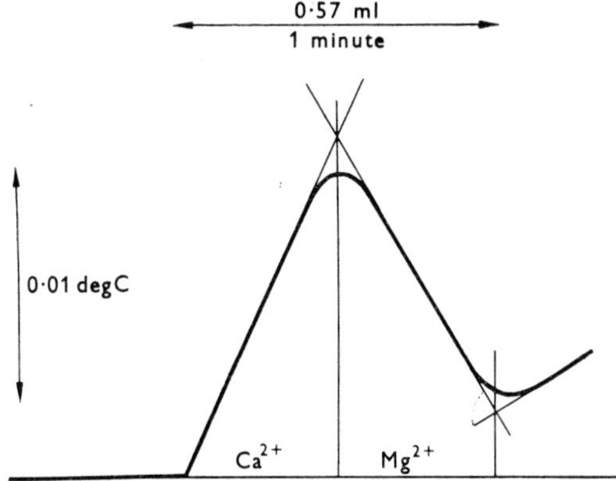

Fig. 3.1 Titration of 5·06 mM Ca^{2+} + 5·40 mM Mg^{2+} with 1·0 M EDTA; from Jordan and Alleman (1957).

Jordan and Alleman (1957) determined calcium with a precision of ±0·4% and an error of 1% and quoted a lowest limit of determination of 1·5 mM with an error of 3%. This latter error is of the same order as that found by Padhye (1957) when he titrated 0·75–3·3 mM Ca, using an indicator method. Eight titrations of M/60 Ca(NO$_3$)$_2$ with 1 M Na$_4$EDTA by automatic digital titration showed a standard deviation of ±1·0% (Priestley, 1963a). Priestley and his co-workers (1968) titrated magnesium sulphate with 1 M Na$_4$EDTA, using a continuous-flow apparatus (see also Section 3.49).

It has been shown that the heat of reaction of calcium with the magnesium/EDTA complex is 25% greater than with Na$_2$H$_2$EDTA and this has been used as a basis for a direct-injection method. Up to 400 mg of CaO in 50 ml of a slightly alkaline solution containing NH$_4$Cl was injected with 10 ml of a neutral solution containing 325 g of Na$_2$H$_2$EDTA and 220 g of MgSO$_4$.7H$_2$O per litre (Danest and Trischler, 1971).

3.6.2 *Determination of calcium as calcium oxalate*

The classical gravimetric method for calcium has also been studied extensively as a thermometric method, because the precipitation is both rapid and exothermic. The presence of magnesium complicates the precipitation because of the post-precipitation of magnesium oxalate. This precipitation has been studied by Brescia and Peisach (1954) who showed that poly(magnesium oxalates) of the form (MgC$_2$O$_4$)$_n$, where $n = 1$–4, are intermediates in the formation of the precipitate $MgC_2O_4 \cdot 2H_2O$. In the kinetics of the reaction the step from $n = 2$ to $n = 3$ was shown to be very slow, and because of this magnesium has little effect on the thermometric method for calcium, provided the titration is carried out quickly enough. When compared to the iodometric titration and the complexometric titration with EDTA, these methods for the determination of calcium in the presence of magnesium prove to be rapid and their precision and accuracy favourable. It is interesting to note that in the precipitation of calcium oxalate there is a zero entropy change which has been attributed to the hydration of the calcium ions. On the precipitation of the calcium these water molecules of hydration are released, i.e., there is a gain in entropy and this must be equal to the entropy loss in the precipitation of the crystalline calcium oxalate (Tyrrell and Beezer, 1968).

CONTINUOUS TITRATION METHODS

Jordan and Billingham (1961) used an automatic apparatus and titrated 50 ml of 0·00915 M Ca^{2+} solution with 0·24 M $(NH_4)_2C_2O_4$, and also, under the same conditions, 0·00905 M Mg^{2+}. Their results showed that magnesium is not titrated regardless of the fact that the titration was carried out at pH 8, at which the precipitation of $MgC_2O_4.2H_2O$ is much more likely.

The method was shown to be very reliable for the precision and error did not exceed $\pm 0\cdot6\%$ in the concentration range 5–100 mM Ca^{2+}, but below 0·5 mM Ca^{2+} it became unreliable, with an error of 50%. In the presence of magnesium it was found to be less reliable for if the molar ratio of Mg^{2+}/Ca^{2+} exceeded a value of 2·2 no end-point was obtained, as can be seen in Table 3.5.

Table 3.5 Precision and accuracy of the determination of calcium as oxalate in the presence of magnesium

(From Jordan and Billingham (1961))

Taken, mM		$Ca^{2+}]$ found*	Molar ratio	Precision†	Error
Mg^{2+}	Ca^{2+}	mM	Mg^{2+}/Ca^{2+}	$\pm\%$	%
1·099	10·50	10·50	0·1048	0·6	0·0
5·495	10·50	10·52	0·5236	0·9	+0·2
13·21	11·97	12·04	1·064	1·0	+0·7
21·11	9·879	9·958	2·127	1·0	+0·7
42·22	9·878	n.e.	4·166	—	—
10·96	52·95	53·16	0·207	0·6	+0·4
26·30	52·95	52·53	0·497	0·4	−0·8
54·80	52·95	52·60	1·036	0·5	−0·7
105·6	49·40	49·12	2·137	1·4	+0·6
205·6	49·40	n.e.	4·273	—	—

* Mean of 3 results.   n.e. No end-point.   † Standard deviation

Jordan and Billingham used this method to determine calcium in limestone and dolomite. The samples were dissolved in hydrochloric acid, ammonia was then added to precipitate iron, etc., and after cooling the solution was titrated. The results obtained using simulated and actual limestones and dolomites are given in Table 3.6.

Jordan and Billingham reported that it was unnecessary to remove the precipitated hydroxides. This finding is confirmed by the Hungarian direct-injection studies (see below), but filtration was recommended in the early titration method used by Chatterji in the analysis of dolomite.

Using the early titration method, Mayr and Fisch (1929) titrated calcium chloride in neutral or slightly acetic acid solution with 0·5 N $(NH_4)_2C_2O_4$.

Table 3.6 Determination of calcium in simulated and actual limestones and dolomites

(From Jordan and Billingham (1961))

Sample	*Concentrations, known or added*				*Tests*	*Calcium found*	*Precision* ±%	*Error* %
	Mg	Al	Fe	Ca				
Synthetic 1	4·497	1·0	2·0	19·88 mM	3	20·07 mM	1·0	+1·0
Synthetic 2	4·497	1·0	2·0	19·88 mM	3	19·95 mM	0·8*	+0·4
Limestone	2·19	4·16	1·63	41·32%, as oxides	5	40·0% CaO	0·7	−1·0
Limestone	1·94	5·70	1·72	37·65%, as oxides	3	37·2% CaO	1·0	−1·0
Dolomite	21·48	0·07	0·08	30·49%, as oxides	3	30·2% CaO	0·4	−1·0

* No precipitate of hydroxides

They obtained results within ±0·3% with no bias when compared with the gravimetric (CaO) method, but stated that concentrations of less than 0·06 g/60 ml gave poor end-points.

Chatterji (1955), using a similar technique, titrated 0·1 g of calcium or an equivalent quantity from a dolomite sample prepared as above but with the precipitated hydroxides filtered off. The solution containing the calcium was diluted to 250 ml, acidified to pH 4 with acetic acid and titrated with approximately 0·5 N $(NH_4)_2C_2O_4$. The results had an error of ±0·3% with no bias, and are not as precise as those that can be expected from gravimetric methods but even this early titration method has the advantage of speed.

DIRECT-INJECTION METHODS

Hungarian workers (Sajó, 1966b) have used oxalate as precipitating reagent in a direct-injection method for calcium in slags, cements and clinkers (Sajó and Sipos, 1966, 1968a, 1969b), in ores, rocks and industrial silicates (Sajó, 1969c) and in magnesites and dolomites (Sajó and Sipos, 1968d). The samples were prepared as described under Manganese (Section 3.35.1) and 5 ml of saturated potassium permanganate added to prevent interference from manganese; alternatively the solution remaining after the titration of manganese was used (see Section 3.35.1).

The prepared solution was titrated with 8 ml of a nearly saturated solution of potassium oxalate (ammonium oxalate is not sufficiently soluble to be effective in this method). The nearly saturated solution (950 ml of saturated solution diluted to 1 litre) was modified if necessary to give no heat of dilution when 8 ml were injected into a blank solution. However, if a heat of dilution occurred, small amounts of either 5% potassium hydroxide or saturated oxalic acid were added to the titrant dependent upon whether the temperature rose or fell, until no heat of dilution occurred. The results obtained are shown in Table 3.7 and compare well with those obtained by other methods.

Table 3.7 Determination of calcium as CaO in slags, clinkers and cements

(Comparison of methods; from Sajó and Sipos (1966, 1968a))

Sample		CaO, %	
		Complexometric method	*Enthalpimetric method*
Blast-furnace slags	1	43·1	43·8
	2	44·6	44·1
	3	41·9	41·2
Siemens-Martin slags	1	44·2	44·7
	2	40·7	40·7
	3	46·7	47·2
	4	43·6	43·4
Clinkers	1	65·1	65·4
	2	65·9	65·7
	3	66·7*	66·2
	4	65·1*	64·8
Cements	1	59·6*	60·2
	2	58·7*	58·2

* Method not stated

With samples containing less than 10% of calcium Sajó and Sipos (1968d) found that the precipitation rate of the calcium oxalate was too slow to yield reliable results, the precipitation time increasing from less than 1 minute to over 10 minutes with contents of 0·5%. They overcame this difficulty by adding 20 ml of 10% $CaCl_2$ solution. In twelve replicate titrations of a magnesite sample known to contain 3·56% CaO, a range of 3·3–3·7% with an average value of 3·48% was obtained. The results obtained on magnesites and dolomites are shown in Table 3·8.

The solution remaining after the titration of calcium in the Sajó and Sipos methods can be used for the determination of magnesium (see Section 3.5.2).

Russian workers using a similar method have obtained results on samples containing between 1·75% and 3·15% of calcium that were correct within ±0·27% absolute (Lopachak and Gudz, 1971).

Table 3.8 Determination of calcium as % CaO in magnesites and dolomites

(Comparison of results; from Sajó and Sipos (1968d))

Sample	*Known* CaO *content** %	CaO *content found* %
Magnesite O	3·56	3·50
Magnesite K	0·85	0·60
Magnesite VK1	1·64	1·40
Dolomite	52·4	52·8
Burnt dolomite	26·9	27·1

* Method not stated

3.7 Strontium

Priestley (1963a) attempted to titrate strontium nitrate with M Na_4EDTA, using an automatic digital titrator, but without success. The only other published thermometric method for strontium is by precipitation as oxalate.

3.7.1 *Determination as strontium oxalate*

Mayr and Fisch (1929) titrated 0·4396 g of strontium chloride in 150 ml of solution with 0·5 N $(NH_4)_2C_2O_4$, using the early titration method, and obtained a value of 0·4406 g. They did not consider the method to be as accurate for strontium as for calcium.

3.8 Barium

3.8.1 *Determination of barium as the EDTA chelate*

Barium, which is difficult to determine volumetrically, using metallo-chromic indicators (see Sijderius, 1954), can be more easily titrated thermometrically with M Na_4EDTA as the titrant to give a well-defined end-point. Priestley and co-workers (1963) titrated 1·0 mmole, of Ba^{2+} in 20 ml of solution and obtained a T value of 0·09 degC/mmole, using an automatic titrator. They found that ammonium ions, which gave a very poor end-point when titrated alone, considerably increased the sharpness of the barium end-point when added in equimolar proportion.

A relative standard deviation of $\pm 0·9\%$ was obtained in ten replicate titrations of M/60 $BaCl_2$ by automatic digital titration (Priestley, 1963a); up to 0·500 arbitrary concentration units of Ba^{2+} were measured to within 0·13% by continuous-flow titration (Priestley *et al.*, 1965, 1968).

3.8.2 *Determination of barium as sulphate*

This classical precipitation method has been studied as a possible thermometric method (Barthel *et al.*, 1968). Figure 3.2 shows the influence of nitric acid concentration on the thermometric curve that was obtained in the titration of 200 ml of 0·05 M Ba(NO$_3$)$_2$ with 1·0 M Na$_2$SO$_4$ and Table 3.9 gives results that show the effect of various titrants and media on the accuracy of the end-point.

This titration contradicts the principles laid down for the quantitative precipitation of barium sulphate, particularly with respect to temperature and nucleation (Smith, 1952), and this accounts for the poor precisions

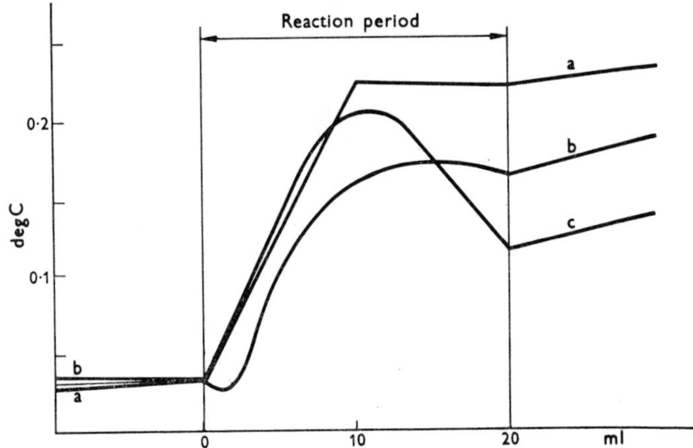

FIG. 3.2 Titration of 10 mmole of Ba(NO$_3$)$_2$ with 1·0 M Na$_2$SO$_4$: in water a; in 0·5 M HNO$_3$ b; in 1·0 M HNO$_3$ c; from Barthel *et al.* (1968).

given in Table 3.9. These difficulties have been overcome by Sajó and Sipos in a direct-injection method in which they added a suspension of barium sulphate to the titrand, prior to the titration, to act as a source of nuclei for the precipitation. The titrant consisted of 100 g of ammonium chloride

Table 3.9 Accuracy of the end-point in the titration of barium with sulphate

(From Barthel *et al.* (1968))

Substance	Titrand Concentration, M	Solvent	Titrant*	Error %
Ba(NO$_3$)$_2$	0·050	water	Na$_2$SO$_4$	0·2
Ba(NO$_3$)$_2$	0·025	water	Na$_2$SO$_4$	0·2
Ba(NO$_3$)$_2$	0·050	3 M NaCl	Na$_2$SO$_4$	1·0
Ba(NO$_3$)$_2$	0·050	0·5 M NHO$_3$	Na$_2$SO$_4$	−7·0
Ba(NO$_3$)$_2$	0·050	water	H$_2$SO$_4$	0·2
BaCl$_2$	0·050	water	H$_2$SO$_4$	0·5
Ba(OH)$_2$	0·025	water	H$_2$SO$_4$	0·2

* All 1 M

in 350 ml of water to which were added 95 ml of concentrated sulphuric acid. The titrant was modified, if necessary, to give no heat of dilution when 5 ml were injected into a blank solution. However, if a heat of dilution occurred small amounts of either ammonium chloride or sulphuric acid were added to the titrant, dependent upon whether the temperature rose or fell, until no heat of dilution occurred.

In the determination of the barium content of slags (Sajó and Sipos, 1966) or of silicates (Sajó, 1969c) the samples were taken into solution with acid or by alkaline fusion. The final solution, which should contain about 50 ml of concentrated hydrochloric acid, was diluted to 200 ml, and 5 ml of barium sulphate suspension (100 g neutral barium sulphate in 400 ml) were added, and when thermal equilibrium was attained 5 ml of the titrant were injected. Table 3.10 gives the results obtained on series of blast-furnace slags, together with those obtained by a gravimetric method and shows that where speed can compensate for loss of precision the direct-injection method should be used.

Table 3.10   Determination of barium as BaO in slags

(Comparison of results from Sajó and Sipos (1966))

Blast-furnace slag	BaO, %	
	Gravimetric method	Enthalpimetric method
1	2·83	2·9
2	2·07	2·2
3	4·46	4·3
4	0·58	0·5
5	0·07	0·0
6	3·26	3·3

The solution remaining after the determination of barium can be used for the determination of iron (see Section 3.36.1(v)).

3.8.3 *Determination of barium as oxalate*

Mayr and Fisch (1929) titrated 0·6845 g of barium chloride in 150 ml of solution with M $(NH_4)_2C_2O_4$, using the early titration method and obtained values of 0·6974 g and 0·6845 g. This poor precision and accuracy was attributed to the unfavourable solubility product.

3.8.4 *Formation of a* $BaCl_2 \cdot KCl$ *complex*

The complex $BaCl_2 \cdot 2KCl$ was shown to be formed with an enthalpy of formation of 1·42 ± 0·02 kcal/mole (Yagub'yan *et al.*, 1965).

Note

The titration of barium acetate is discussed in Section 4.1.4(*ii*).

3.9 Boron

3.9.1 *Determination of boron*

The thermometric determination of boron has been confined to those cases where it is present as boric acid. Boric acid can be easily titrated thermometrically with alkali and this is discussed under Neutralization Reactions (Section 3.48), but it can also be determined as the borofluoride or mannitol complexes. That these reactions can be utilized analytically gives an idea of the versatility of the technique.

(i) DETERMINATION OF BORATE AS THE BOROFLUORIDE COMPLEX

Boric acid reacts exothermally and rapidly with hydrofluoric acid to form the borofluoride complex, but Sajó and Sipos (1967a), developing a direct-injection method based on this reaction, found that the 40% hydrofluoric acid titrant had a large heat of dilution. They overcame this difficulty by adding urea to give a titrant with no heat of dilution (see Fig. 2.34).

The method was used to determine the boric acid content of plating solutions, and a solution of 205 g of urea in 205 ml of 40% hydrofluoric acid was used as the titrant which was expected to show no heat of dilution when 5 ml, contained in a platinum crucible, were added to 200 ml of a blank solution. If the solution temperature rose or fell, urea or 40% hydrofluoric acid was added to the titrant until it showed no heat of dilution. Determinations of between 10 and 50 g of boric acid per litre in the presence of nickel, sulphate and chloride agreed to well within $\pm 0{\cdot}6$ g/l.

It has been suggested that boric acid can be determined by a displacement reaction in which boric acid is titrated into complex metal fluoride solutions with the formation of the borofluoride complex. Complex fluoride solutions containing aluminium, iron or antimony fluorides were titrated and the latter gave a titration curve showing a sharp inflection (Deschamps *et al.*, 1968).

(ii) DETERMINATION OF BORATE AS THE MANNITOL COMPLEX

The borate content of silicate solutions has been determined as the mannitol complex. The solutions were made alkaline with potassium hydroxide solution and diluted to 200 ml after the addition of ammonium chloride as a buffer. Saturated mannitol solution was then injected into the solutions (Sajó, 1969c).

3.9.2 *Boron trifluoride as a Lewis acid*

The heat of reaction of boron trifluoride acting as a Lewis acid in benzene solution with the Lewis bases anisole and phenetole has been determined by thermometric titration (Romm *et al.*, 1968).

3.10 Aluminium

3.10.1 *Determination of aluminium as the* EDTA *chelate*

Aluminium has been titrated thermometrically, using automatic titrators, the reaction being endothermic. A relative standard deviation of $\pm 0{\cdot}8\%$ was obtained in eight replicate titrations of $M/60$ $AlCl_3$ with M Na_4EDTA in an automatic digital titrator (Priestley, 1963a). The precision is very similar to that obtained by an indicator titration method

(Taylor, 1955). In a titration of 0·3 mmole of Al^{3+} in 25 ml of solution in the presence of iron, calcium, magnesium and barium (Priestley *et al.*, 1963) the end-point was not stoichiometric, owing to substitution reactions taking place ($T = 0.13$ degC/mmole). Jordan and co-workers (1966) suggested that micro-injection of EDTA solution could be used to determine aluminium in the presence of lead, copper, nickel, indium, manganese or cobalt.

3.10.2 *Determination of aluminium as* H_3AlF_6

The reaction of aluminium ions with hydrofluoric acid to form AlF_6^{3-} ions takes place with the evolution of heat, but the reaction is very slow. Sajó and Sipos found that phosphoric acid accelerated the reaction and used this as the basis of a direct-injection method for the determination of the aluminium content of blast-furnace slags (1966) and of cements and clinkers (1968a).

In the method, silica which interferes was first precipitated and dehydrated with perchloric acid and/or sulphuric acid and filtered off. The filtrate was received in a plastic beaker, phosphoric and hydrochloric acids were added to it, and the mixture was diluted to a known volume. The titrant, injected by means of a 5-ml platinum pipette or crucible, consisted of a solution of 150 g of urea in 40% hydrofluoric acid and was modified, if necessary, to give no heat of dilution when 5 ml were injected into a blank solution containing the reagents. If a heat of dilution occurred, small amounts of either urea or hydrofluoric acid were added to the titrant dependent upon whether the temperature rose or fell, until no heat of dilution occurred.

There was interference from titanium(IV) and iron(III); the latter is not important in blast-furnace slags but can be allowed for by the small correction, 1.00% $Fe_2O_3 \equiv 0.03\%$ Al_2O_3. The interference from titanium, which is common to other methods for the determination of aluminium, is more serious, but in this method can be allowed for by the use of the correction, 1.00% $TiO_2 \equiv 0.78\%$ Al_2O_3, or, if hydrogen peroxide is present (see Section 3.32.2(*iii*)) 1.00% $TiO_2 \equiv 0.48\%$ Al_2O_3. The results obtained by the method are shown in Table 3.11 and as can be seen do not compare at all favourably in precision with those obtained by a complexometric indicator method or even with results obtained by an atomic-absorption method which is equally rapid but has been shown to have a repeatability of 0·22% at this level of Al_2O_3 (Cobb and Harrison, 1971).

The method has also been used to determine the aluminium content of aluminate solutions (Sajó *et al.*, 1969). The results are shown in Table 3.12.

3.10.3 *Determination of aluminium as* NaK_2AlF_6

Aluminium in solution in hydrofluoric acid to which potassium chloride has been added exists in the form K_3AlF_6. If sodium chloride is added the compound elpasolite, NaK_2AlF_6, is precipitated with the evolution of heat. This reaction has been utilized to determine aluminium in silicates (Sajó, 1969c), in clays (Sajó and Sipos, 1968b), and in magnesites and dolomites (Sajó and Sipos, 1968d). The analytical use of this reaction is unique to enthalpimetric titrimetry and has also been utilized in the determination of sodium (see Section 3.2.1). The reaction was carried out in a hydrofluoric/hydrochloric acid solution or alternatively the solution after the titration of silicon as $(SiF_6)^{2-}$ or as K_2SiF_6 was used (see Section 3.15.1 or 3.15.2).

Table 3.11 Determination of aluminium as % Al_2O_3 in slags, clinkers, and cements

(Comparison of methods; from Sajó and Sipos (1966, 1968a))

Sample		Complexometric method	Enthalpimetric method
Blast-furnace slags	1	8·67	8·8
	2	6·50	6·2
	3	8·28	8·5
Siemens-Martin slags	1	3·57	3·7
	2	4·55	4·3
	3	3·17	3·4
	4	4·08	4·3
Clinkers	1	6·67	6·9
	2	6·03	6·4
	3	7·33*	7·2
	4	7·54*	7·7
Cements	1	7·26*	7·3
	2	6·72*	6·9

* Method not stated

Table 3.12 Determination of aluminium as Al_2O_3 (g/l) in aluminate solutions

(Comparison of methods; from Sajó et al. (1969))

Sample		Complexometric method	Enthalpimetric method
Technical aluminate	1	147	148
	2	126	126
	3	112	113
	4	98	98
	5	50	50
Synthetic aluminate	1	192	191
	2	96	96
	3	48	50
	4	24	23

Titanium does not appear to interfere with this method but calcium, if present above 2%, interferes and under these circumstances the $KCaAlF_6$ precipitation method can be used (see the next section). Alternatively the hydroxides precipitated by ammonia, for example, from the determination of manganese, calcium or magnesium (see Section 3.6.2(i)), can be filtered off, washed and dissolved in hydrochloric/hydrofluoric acid. In all cases 2 ml of 25% KCl solution were added to precipitate silicon as K_2SiF_6, and, after thermal equilibrium was established, 2 ml of the titrant were added. The titrant consisted of a 25% solution of sodium chloride to which 2% of triethanolamine had been added. The titrant was modified if necessary to give no heat of dilution when 2 ml were injected into a blank solution; if the temperature fell small amounts of triethanolamine were added to the titrant and if the temperature rose 2 M HCl was added until no heat of dilution occurred.

A series of twelve replicate determinations analysing the precipitated hydroxides from a magnesite known to contain 0·21% Al_2O_3 had a range of 0·16–0·30% with an average of 0·23%. The results of the determination of aluminium in clays, magnesite and dolomites are given in Table 3.13 and it can be seen that there is good agreement, particularly at the higher levels, with the known contents.

Table 3.13 Determination of aluminium as % Al_2O_3 in clays, ores, magnesites and dolomites

(Comparison of methods; from Sajó and Sipos (1968b, d))

| | | Content found | |
Sample	Known content	Direct	Separated M(OH)$_3$
Fireclay 1	37·4	37·1	—
Fireclay 2	34·7	34·2	—
Fireclay 3	54·7	54·2	—
China clay	34·4	34·5	—
Felspar	18·4	18·0	—
Granite	14·3	14·2	—
Magnesite O	0·21	—	0·16
Magnesite K	0·48	—	0·40
Magnesite VK1	0·37	—	0·50
Dolomite	0·27	—	0·19
Burnt dolomite	0·41	—	0·50

The solution remaining after the determination of aluminium can be used for the determination of iron (see Section 3.36.2(i)).

A variant of this procedure, in which a thermometric method was used, was developed by Everson (1971) who used potassium fluoride as the titrant. A sample containing between 5 and 40 mg of aluminium was taken and 0·5 g of ascorbic acid added to mask iron and titanium. Using a meter, the pH was adjusted to 2·0–2·2 and 5 ml of reagent solution (160 g of NaCl + 100 g of NaOOCCH$_3$ + 80 g of K$_2$SO$_4$ in 1 litre) were added and the pH again adjusted to 3·9 ± 0·2 by the addition of saturated sodium acetate. The final solution, diluted to 40 ml, was titrated with 2 M KF. As with the direct-injection method, calcium interfered, but beryllium, zirconium and thallium were also stated to interfere. The method was used to analyse a number of U.S. Bureau of Standard Samples materials and the results obtained were precise. For example, a Ni-Cu alloy containing 0·23% Al gave a result of 0·224% and a burnt refractory containing 37·67% Al$_2$O$_3$ gave results of 37·58%, 37·85%, 37·38% and 37·40%.

3.10.4 *Determination of aluminium as KCaAlF$_6$*

Calcium, if present above 2%, interfered with the NaK$_2$AlF$_6$ precipitation method above because the compound KCaAlF$_6$ was precipitated with the evolution of heat (Sajó, 1969c). If large amounts of calcium were present in the sample (or presumably added to it) it was found that after the initial heat rise on the injection of hydrofluoric acid to the sample solution (as in the determination of silica as K$_2$SiF$_6$; see Section 3.15.2) the temperature did not change for a few minutes, then it rose suddenly because of the precipitation of KCaAlF$_6$, the delay being caused by the slow formation of crystal nuclei. The temperature rise was found to be proportional to the aluminium content and the time for the determination including that of the silica content was 20 minutes.

3.10.5 *Reaction of aluminium salts with fluoride*

Sodium fluoride solutions have been titrated with aluminium chloride; both at various concentrations. All the titration curves except those obtained with the lowest concentrations (about 0·1 N) showed sharp inflections due to the formation of $(AlF_6)^{3-}$ ions (Deschamps *et al.*, 1968). Titrations of a mixture of iron and aluminium chlorides with sodium fluoride did not distinguish between the two chlorides. It was also shown that if the $(AlF_6)^{3-}$ ions were titrated with boric acid the titration curve had an inflection due to the formation of the aluminium fluoroborate.

3.10.6 *Reaction of aluminium(III) with alkalis*

(*i*) DETERMINATION WITH CARBONATE

Using an automatic digital titrator, Priestley (1963a) obtained a relative standard deviation of ±3·0% in six replicate titrations of M/90 AlCl$_3$ with M Na$_2$CO$_3$. The reaction in this case was endothermic.

(*ii*) REACTION WITH SODIUM HYDROXIDE

Dilute or concentrated solutions of aluminium salts have been titrated with sodium hydroxide solution by the early titration method (Grobet, 1921). A number of inflections occurred in the titration curves, from which the composition of the basic complexes of aluminium have been determined as given in Table 3.14. One of the inflections was sharp enough to be suitable for a precise analysis.

Table 3.14 The composition of the basic complex salts of aluminium
(From Grobet (1921))

Aluminium compounds	NaOH titrant	Successive complexes formed
Nitrate	Dilute	$Al(OH)_3$; $AlO(ONa)$; $Al(ONa)_3$
Chloride; Sulphate; Potassium alum	Dilute	$Al(OH)_3$; $Al(ONa)_3 \cdot Al(OH)_3$; $Al(ONa)_3$
Nitrate; Chloride; Sulphate	Conc.	$Al(X^-)_3 \cdot Al(OH)_3$ *or* $Al_2(Y^{2-})_3 \cdot 2Al(OH)_3$; $Al(OH)_3$; $AlO(ONa)$; $Al(ONa)_3$
Potassium alum	Conc.	$Al_2(SO_4)_3 \cdot 2Al(OH)_3$; $Al(OH)_3$; $Al(ONa)_3$

Barthel and co-workers (1968) titrated 200 ml of 0·025 M $AlK(SO_4)_2 \cdot 12H_2O$ with 2M NaOH and obtained an error of 1% ($T = 0·99$ degC/mmole). From the titration curve they calculated the heat of precipitation of $Al(OH)_3$ as $-5·1 \pm 0·2$ kcal/mole and the $\Delta H°$ of the reaction $Al(OH)_3 + OH^- \rightleftharpoons [Al(OH)_4]^-$ as $+1·1 \pm 0·2$ kcal/mole.

3.10.7 *Formation of organic acid chelates of aluminium*

The oxalate, malonate, tartrate and citrate complexes of aluminium have been investigated by Gallet and Pâris (1967a, b). The titrand contained potassium chloride, or in the case of the citrate titration, ammonium chloride, in order to increase its ionic concentration. A summary of the results that were obtained is given in Table 3.15. The 1 : 1 Al^{3+} tartrate and citrate chelates were also found by Cadariv and Goina (1961) who reported that the reaction of sodium aluminate with tartrate was exothermic.

Table 3.15 The composition of the organic acid chelates of aluminium
(From Gallet and Pâris (1967a, b))

Titrand	Titrant	Chelates formed, ratio Al : Acid
Oxalate, 1·3518 M, pH 7	$Al(NO_3)_3$ 1·008 M, pH 2	1 : 3, 1 : 2
$Al(NO_3)_3$, 1·002 M, pH 4	Malonate 3·002 M, pH 6·5	1 : 1, 1 : 2, 1 : 3
$Al(NO_3)_3$, 0·5025 M, pH 4	Tartrate, 2·042 M, pH 10	1 : 1, 2 : 3, 1 : 2, 4 : 11, 1 : 4
$Al(NO_3)_3$, 0·5021 M, pH 4	Citrate, 1·014 M, pH 10	3 : 2, 1 : 1, 2 : 3, 1 : 2

3.10.8 *Reaction of aluminium metal with acid*

The solution rate of metals in acid has been found to be dependent on the concentration of impurities present in the metal. One method for the determination of the dissolution rate is to measure the temperature of the acid during the reaction and this has been used to determine the purity of aluminium.

Mylius (1922), in an early method, used a modified test-tube in which were placed 20 ml of 10% hydrochloric acid. A 0–100 °C thermometer was suspended in the acid. Standard-sized cleaned aluminium wire was added to the acid and the maximum temperature $(T°)$ and time taken (min.) for this to be reached was noted. The value of $T°$ was stated to be a function of the reactivity or corrosion index of the aluminium and the reaction class was given by the equation $(T - 20)/\text{min}$.

Sajó and Sipos (1967b) improved the method by using a Dewar flask in which were placed 200 ml of 10% hydrochloric acid together with a stirrer and thermistor. The sample in the form of a cleaned wire was suspended in the acid by a synthetic thread. The sensitivity of the bridge network connected to the thermistor was adjusted, the first half-minute of reaction time being used as a guide. After one minute's reaction the temperatures were read at half-minute intervals for about three minutes. An even temperature rise was generally found over several half-minutes; this steady rate was noted and from it the concentration of impurities was read from a calibration graph. The calibration graph was prepared by plotting the logarithms of the rates against the logarithms of the concentration of the impurities in standard aluminium samples. For very pure aluminium ($\geqslant 99.9\%$), the reaction rate was very slow and a smaller volume of concentrated hydrochloric acid was used to increase the rate.

In all the methods the samples of wire were cleaned with sodium hydroxide followed by hydrochloric acid, and washed with distilled water before and after each stage.

3.10.9 *Lewis acid activity of aluminium compounds*

(*i*) ALUMINIUM TRICHLORIDE

The Lewis acid activity of aluminium trichloride was determined by Trambouze and Pascal (1951), using the early titration method. Aluminium trichloride suspended in benzene was titrated with either 1 M or 2 M benzene solutions of dioxan or ethyl acetate. A 2–5 degC temperature rise was observed and the error was of the order of 1%. The authors stated that the method was of value in ascertaining the formation of intermediates in the Friedel-Craft reaction (see also Section 4.4, page 219).

(*ii*) ALUMINIUM TRIBROMIDE

The heats of reaction of aluminium tribromide with various ethers acting as Lewis bases have been determined in benzene solution. An average ΔH value of -23.3 kcal/mole was obtained for the diethyl, dipropyl, dibutyl and dioctyl ethers. Other ethers studied were anisole, phenetole, diphenyl ether, di-*p*-tolyl ether, di-*p*-bromophenyl ether, 2-methoxy- and 1-ethoxy-naphthalene. All the compounds formed were of the AB type and had a smaller ΔH with a value of approximately −15 kcal/mole (Romm *et al.*, 1967).

(*iii*) Aluminosilica catalysts

The Lewis-acid active centres of aluminosilica catalysts have been determined by suspending the finely ground catalyst in benzene and titrating it with a Lewis base. Trambouze and co-workers (1952), using the early titration method, used dioxan and ethyl acetate as titrants. In later work Valcha (1965) described the titration of 10 g of sample in 40 ml of benzene with 1 M ethyl acetate. The results showed that there was considerable variation between samples. The anthraquinone or dimethyl yellow indicator volumetric method with n-butylamine as the titrant gave lower results than the thermometric method.

Note

For the determination and titration of alkyl aluminium compounds see Section 3.50; for the determination of the composition of aluminium arsenates see Section 3.20; for the determination of the composition of aluminium fluorides see Section 3.26.1, and for the determination of small amounts of aluminium by a catalytic method see Section 3.52.

3.11 Gallium

3.11.1 *Formation of organic acid chelates of gallium*(III)

The malonate, tartrate and citrate chelates of gallium(III) have been investigated by titration of gallium(III) chloride with solutions of the acids (Gallet and Pâris, 1967a, b). The gallium solution contained potassium nitrate to increase the ionic strength and the results obtained are summarized in Table 3.16.

Table 3.16 Composition of organic acid chelates of gallium(III)
(From Gallet and Pâris (1967a, b))

Gallium(III) *chloride*	*Titrant*	*Chelates formed* Ga : *Acid*
0·3041 M, pH 2	Malonate, 2·721 M, pH 6	1 : 1, 1 : 3
0·3041 M, pH 2	Tartrate, 1·425 M, pH 10	1 : 1, 1 : 2, 1 : 5
0·6731 M, pH 2	Citrate, 1·042 M, pH 11	3 : 2, 1 : 1, 2 : 3, 1 : 2

Note

For the determination of small amounts of gallium by a catalytic method see Section 3.52.

3.12 Indium

The only reference to a thermometric method for the determination of indium is the catalytic one described in Section 3.52.

3.13 Thallium

3.13.1 *Formation of thallium(I) hexacyanoferrate(II)*

The early titration method has been used to determine the composition of this thallium compound (Gaur *et al.*, 1958). The compound $Tl_4[Fe(CN)_6]$ was formed when 0·2 M $K_4[Fe(CN)_6]$ was titrated with 0·2 M $TlNO_3$.

3.13.2 *Thallium(III) halide complexes*

A study of the complexing of thallium(III) with Cl^- and Br^- has shown that only mononuclear complexes are formed and that not more than four bromide ions can be complexed by a thallium(III) ion. The reaction was carried out in M $NaClO_4$ with M $NaCl$ or $NaBr$, and the thermodynamic properties of the reactions were determined (Léden and Ryhl, 1964).

Note

For the determination of small amounts of thallium by a catalytic method see Section 3.52. It should be possible to use tetraphenylborate as a titrant in the determination of thallium (see Section 3.3.2).

3.14 Carbon

3.14.1 *Determination of carbon dioxide*

The technique of gas-injection enthalpimetry developed by Zambonin and Jordan (1969) was shown to be ideally suited to the determination of carbon dioxide in a gas by injection into an excess of 0·5 M KOH containing 0·2 M K_2CO_3. The relevant kinetics are fast, the equilibrium virtually complete with no side-reactions, and the heat of reaction is high ($\Delta H = -27$ kcal/mole). On the small scale as little as 0·5 μmole of CO_2 could be determined but the lowest level of carbon dioxide that could be determined with an error and precision of better than $\pm 2\%$ was 10 ml/l.

A disadvantage of the method is that the sample has to have the same water-vapour pressure as that of the potassium hydroxide solution, otherwise a blank injection has to be made of a carbon-dioxide free gas of the same water-vapour pressure as the sample. Sulphur dioxide and nitric oxide will obviously interfere.

3.14.2 *Determination of carbonates*

The determination of carbonates by titration with acid is discussed in Section 3.48.2(*iii*) and the only reference to date to other methods for the determination of carbonates is the suggestion by Priestley and his co-workers (1963) that magnesium chloride might be used as a titrant for the determination of carbonate in admixture with bicarbonate because magnesium hydrogen carbonate is sufficiently soluble and should not therefore interfere with the carbonate end-point.

3.14.3 *Determination of carbon disulphide*

Bluck (1970) determined traces of carbon disulphide in hydrocarbons by titration with an alcoholic solution equimolar in both potassium hydroxide and dimethylamine. The reaction was extremely exothermic for not only is the primary reaction of the carbon disulphide to form diethyl dithiocarbamate itself very exothermic but the neutralization of the product with the alkali also contributes to the heat evolved. All this, combined with the low specific heat of the medium, makes it an extremely sensitive method for carbon disulphide.

3.14.4 *Determination and composition of cyanides*

(*i*) DETERMINATION WITH SILVER NITRATE

Mayr and Fisch (1929), using a Beckmann thermometer in the early titration method, attempted to distinguish between cyanide and chloride in a mixture by titrating with silver nitrate but obtained only a single sharp end-point representing the sum. In the titration of cyanide alone they again obtained only one end-point. Harries (1966), using much more sensitive thermopile temperature-detection, obtained in a cyanide titration a curve showing two inflections corresponding to $[Ag(CN)_2]^-$ and $AgCN$. No analytical results were given in either paper.

(*ii*) COMPOSITION OF CYANIDES

The pK, ΔH°, and ΔS° values for the dissociation of HCN in aqueous solution have been determined as a function of temperature in a study of metal cyanide co-ordination (Izatt *et al.*, 1962).

The compositions of the complex cyanides of iron, nickel, zinc and cobalt have been determined and are given in the appropriate sections.

3.14.5 *Determination of thiocyanate*

Takeuchi and Yamazaki (1969a) titrated a mixture of iodide and thiocyanate with silver nitrate, using a differential titration apparatus, and the differential as distinct from the direct titration curve showed two sharp end-points representing the iodide and the thiocyanate, each of which could be determined with an error and precision of between 1 and 3% for solutions containing 0·01 mole of each per litre. Harries (1966), using a thermopile and burette apparatus, also successfully titrated thiocyanate with silver nitrate as the titrant, but could not use mercurous nitrate as the titrant because of the disproportionation of the mercurous compounds that were formed initially. Keily and Hume (1964) reported that, in a glacial acetic acid medium, thiocyanates could be determined by titration with perchloric acid.

Note

In studies of the measurement of rapid reactions in solution by continuous flow mixing, Roughton (1963) determined the rate of combination of carbon dioxide with sodium hydroxide and with ammonia and of carbon monoxide with haemoglobin, and Chipperfield (1966) determined the rate of combination of carbon dioxide with the anions of amino-acids. For the determination of small amounts of cyanide and of thiocyanate by a catalytic method see Section 3.52.

3.15 Silicon

3.15.1 *Determination of silicon as* H_2SiF_6

The first description of the important Hungarian work on the direct-injection enthalpimetric determination of silicon was in a paper describing the sequential determination of the basicity, iron(II) and silicon content of slags (Sajó, 1957). The solution of the slag, after the determination of the basicity by dissolution in hydrochloric acid, was injected with hydrogen peroxide to determine iron(II), followed by hydrofluoric acid which caused a temperature rise during 2–3 minutes that was proportional to the silicon content to within $\pm2\cdot5\%$. Later work (Mandl *et al.*, 1962) criticized the order of injection of the two reagents because of secondary reactions (see Section 3.36.1(*iv*)).

The reason for the temperature rise is the exothermic reaction $SiO_2 + 6HF \rightarrow H_2SiF_6$. This reaction must be carried out in hydrochloric acid solution, but the silica must be in solution or colloidal solution

Table 3.17 Determination of silicon in standard steels

(Comparison with known contents; from Sajó (1968b))

Standard	Known content %	Found %
SM17	1·21	1·22
SM23	0·37	0·37
SM25	0·20	0·19
SM49	0·10	0·10
SM53	1·46	1·44
SM56	0·73	0·74
SM55	0·76	0·76
SM56	0·70	0·71

for the reaction to be rapid enough for the enthalpimetric method to be effective. The 40% HF titrant is used in excess so that the reaction goes to completion, and because of this precautions have to be taken to prevent reaction with the apparatus. A platinum or plastic injection pipette must be used and the thermistor, Dewar and stirrer must be coated with plastic material.

The difficulty with the high heat of dilution found with the hydrofluoric acid titrant has been overcome in three ways. In one method (Sajó and Ujavari, 1964) a second injection was made which gave a temperature rise caused by the exothermic heat of dilution. This rise was subtracted from that obtained in the first injection caused by the heat of silica reaction and of dilution. In a second method (Sajó, 1968b) differential dual injection to counteract the heat of dilution was used. This method for the determination of the silicon content of standard steels gave the results shown in Table 3.17, which indicate that it is accurate.

In the third method, urea was added to counteract the heat of dilution of the hydrofluoric acid as it has an endothermic heat of dilution and does

Table 3.18 Determination of silicon as % SiO$_2$ in slags, clinkers, cements, magnesites and dolomites
(Comparison with known contents; from Sajó and Sipos
(1966; 1968a, d))

Sample		Gravimetric method	Thermometric method
Blast-furnace slags	1	37·4	37·8
	2	39·1	39·8
	3	33·4	33·6
Siemens-Martin slags	1	17·4	17·1
	2	12·8	12·7
	3	39·0	38·3
	4	21·1	21·4
	5	12·1	12·4
Clinkers	1	21·5	21·7
	2	20·6	20·5
	3	20·9	21·2
	4	21·7*	21·3
Cements	1	22·6*	22·1
	2	23·8*	24·1
Magnesites	O	1·64*	1·68
	K	1·42*	1·50
	VK	0·82*	0·88
Dolomite		3·90*	3·78
Burnt dolomite		0·95*	1·00

* Method not stated

not react with the hydrogen fluoride (Sajó, 1966b). The titrant in this case consisted of a solution of 150 g of urea in 1 litre of 40% hydrofluoric acid, which was modified if necessary to give no heat of dilution when 5 ml were injected into a blank solution, small amounts of urea or hydrofluoric acid being added to the titrant dependent upon whether the temperature rose or fell, until there was no heat of dilution (see Fig. 2.34).

When the titrant with zero heat of dilution was used, a series of twelve replicate determinations of the silicon content of a magnesite sample known to contain 1·64% SiO$_2$ had a range of 1·53–1·70% with an average value of 1·63%.

The results obtained with samples of magnesites and dolomites (Sajó and Sipos, 1968d), of clinkers and cements (Sajó and Sipos, 1968c, 1969b), and of slags and clinkers (Sajó and Sipos, 1966) are given in Table 3.18; they again indicate that the method is precise and where applicable is preferable to the slower gravimetric method.

Interference with these methods is encountered with iron(II) and titanium(III) which react with the hydrofluoric acid, but this can be avoided by oxidizing directly or in the course of the enthalpimetric determination of iron(II) with persulphate (see Section 3.36.1(v)) followed by the determination of titanium(III) with hydrogen peroxide (see Section 3.32.2(ii)). Alternatively, if the content of interfering metals is known, the apparent silicon content can be corrected by using correction factors (Sajó and Sipos, 1966).

The method cannot be used if the sample is taken into solution with hydrofluoric acid, unless the latter is removed again. The silicon content can however be determined under these circumstances by the following method (Section 3.15.2). The solution remaining after the silicon determination can be used for the determination of aluminium as K_2NaAlF_6 (see Section 3.10.3) or of manganese (see Section 3.35.1).

3.15.2 *Determination of silicon as* K_2SiF_6

As an alternative to the H_2SiF_6 method above, if hydrofluoric acid has been used to take the sample into solution, the silicon can be determined by injecting potassium chloride, when the compound K_2SiF_6 is formed with the evolution of heat (Sajó and Sipos, 1968b). A solution of 250 g of potassium chloride in 800 ml of water is used as the titrant, and so that there is no heat of dilution when this is injected the composition of the acids used to dissolve the sample (14 ml of 40% HF + 20 ml of 1 : 1 HCl) is adjusted in a blank determination. Table 3.19 gives the results obtained when using the method to determine the silicon content of various types of clays and minerals and it can be seen that the method is accurate at these high silicon contents.

Table 3.19 Determination of silicon as % SiO_2 in clays and minerals
(Comparison of results; from Sajó and Sipos (1968b))

Sample	*Known content*	*Found*
Fireclay 1	55·6	55·9
Fireclay 2	58·5	58·1
Fireclay 3	39·3	39·5
China Clay	52·3	52·7
Felspar	66·2	66·8
Granite	73·5	73·7

Note

For the determination of active sites on aluminosilica catalysts see Section 3.10.9 and for the reactions of silicotungstates see Section 3.45.2.

3.16 Tin

3.16.1 *Determination of tin(II) by oxidation*

Barthel and Schmahl (1965), using a precise thermometric titration apparatus, have titrated 0·005 M $SnSO_4$ in 1 N H_2SO_4 solution with both potassium permanganate and ceric sulphate. The results agreed within ±1% and had no error. The high $\Delta H°$ value (-36 ± 2 kcal/eq) of the permanganate oxidation reaction, was determined at the same time.

3.16.2 *Reaction of tin with fluoride*

(*i*) FORMATION OF TIN(II) COMPLEXES

The thermometric curve in the titration of tin(II) chloride and sodium fluoride had an inflection attributed to SnFCl but not to SnF_3^- (Deschamps *et al.*, 1968).

(*ii*) DETERMINATION OF TIN(IV) AS $[SnF_6]^{2-}$

Sajó and Sipos (1968c) have utilized the endothermic reaction of tin(IV) ions with fluoride ions to determine the tin content of alloys by the direct-injection method. The sample (0·3 g) was dissolved in hydrochloric acid and the tin oxidized to the stannic stage with potassium chlorate. The solution was then diluted to 200 ml and 8 ml of hydrofluoric acid were injected. The titrant, which had no heat of dilution, was prepared by dissolving approximately 320 g of urea in 1 litre of 40% HF, the amount depending upon whether there was a temperature rise when 8 ml of the mixture were injected into a blank solution. The results obtained on samples of tin alloys of known content are given in Table 3.20 and show good agreement with the known contents.

Table 3.20 Determination of tin in alloys
(Comparison with known contents; from Sajó and Sipos (1968c))

Batch number	Known content %	Found %
GMS 67	1·13	1·10
GZ-RG5	6·32	6·40
LG PbSn10	10·25	9·90
L Sn30	29·3	29·8
LG Sn30/70	29·6	29·4
LG Sn40/60	38·8	39·2
LG Sn50/50	49·7	49·3
LG Sn60/40	61·3	61·0
LG Sn80	79·9	80·2

3.16.3 *Determination of tin(IV) as the* EDTA *chelate*

Priestley (1963a), using an automatic digital titrator, titrated M/120 $SnCl_4$ with 1 M Na_4EDTA and in ten replicates obtained a relative standard deviation of $\pm 0.6\%$.

3.16.4 *Reaction of tin chlorides*

(*i*) TIN(II) CHLORIDE

It has been suggested that tin(II) chloride can be titrated with alkali to the $Sn(OH)_2$ end-point, but the content of free hydrochloric acid should be known (Barthel *et al.*, 1968).

(*ii*) TIN(IV) CHLORIDE

Tin(IV) chloride in non-aqueous aprotic solvents acts as a Lewis acid (A), and by means of thermometric titrimetry it is possible to determine the type of product formed (AB_n) when it reacts with a Lewis base (B) and also to determine tin(IV) chloride by using Lewis bases. The reactions that have been studied are given in Table 3.21 and some of these have been utilized in the quantitative determination of tin(IV) chloride.

Table 3.21 Tin(IV) chloride–Lewis base reactions studied by thermometric titrimetry

Lewis base (B)	*Type formed*	*Reference*
Tetrahydrofuran	AB_2	Zenchelsky and Segatto (1958)
Pyridine	AB_2	,,
1,4-Dioxan	AB	,,
Morpholine	AB	,,
Tetrahydrofuran	AB_2	Cioffi and Zenchelsky (1963)
Tetrahydropyran (THP)	AB_2	,,
2-Methyl THP	AB_2	,,
2,5-Dimethyl THP	AB_2	,,
4-Methyl THP	AB_2	,,
Diethyl ether	AB_2	Gol'dshtein *et al.* (1965b)
$BuS(CH_2)SBu$	AB	,,
$EtO(CH_2)OEt$	AB	,,
R_2S	AB	Gol'dshtein *et al.* (1965a)
$RS(CH_2)_nSR$	AB	,,
$R_nSn(OR)_{4-n}$*	AB	Gol'dshtein *et al.* (1967a)
$MeOSnBu_3$*	AB	,,
$(MeO)_2SnBu_2$*	AB	,,

* The sulphur analogues did not form products

It has been shown by Zenchelsky and his co-workers (1956) that tin(IV) chloride can be titrated in 100 ml of solvent with dioxan in the same solvent.

Results agreeing to within $\pm1\%$ were obtained with 3–35 mmole of $SnCl_4$ in benzene or carbon tetrachloride solution, the upper limit being set by the precipitation of the adduct on the transistor. A poorer precision was obtained in nitrobenzene or chloroform solution. In benzene the heat of reaction was found to be $-16\cdot2$ kcal/mole. The Lewis bases pyridine, tetrahydrofuran and morpholine have also been used to determine tin(IV) chloride in benzene solution (Zenchelsky and Segatto, 1957a).

(*iii*) Organo-tin(IV) chlorides

The Lewis-base activity of organo-tin(IV) chlorides was also studied. It was found that although the compounds R_2SnCl_2 react with $R_nSn(OR)_{4-n}$ bases to form AB compounds, the analogous compounds R_3SnSR and $R_2Sn(SR)_2$ react differently and form compounds of the type $R_2SN(SR)Cl$ (Gol'dshtein *et al.*, 1967a). The compound Bu_2SnCl_2 was found to form AB-type compounds with the bases $MeOSnBu_3$ and $(MeO)_2SnBu_2$ but not with their sulphur analogues (Gol'dshtein *et al.*, 1967b); with the bases $Bu_2Sn(OMe)_2$ and $Bu_2Sn(OBu)_2$ it formed AB compounds (Gol'dshtein *et al.*, 1965a).

Note

For the determination of small amounts of tin by a catalytic method see Section 3.52.

3.17 Lead

3.17.1 *Determination of lead as the* EDTA *chelate*

Jordan and Alleman (1957) showed that the heat of chelation of lead(II) with EDTA was high ($\Delta H = -12\cdot8$ kcal/mole) and in the same titration of $0\cdot01$ M Pb^{2+} with M Na_4EDTA obtained a precision and error of $\pm1\%$ and quoted a lowest limit of detection of $0\cdot5$ mmole/1, with an error of $\pm3\%$. This precision is the same as that obtained by Pinkston (1955) in an indicator titration method. The direct-injection method has also been used with approximately 1 M Na_4EDTA as titrant. Wasilewski and co-workers (1964) injected 300 μl into 25 ml of solution containing between $49\cdot4$ and 201 μmole of Pb^{2+} and in this very rapid method obtained results that agree to within $\pm3\%$. Jordan and co-workers (1966), using a microtechnique, injected 40 μl into 2 ml containing 2–40 mmole of $Pb(NO_3)_2$ and obtained a linear calibration graph. Five replicate titrations had a precision of $\pm1\%$ and an error of less than $\pm2\%$. Between 200 and 700 arbitrary concentration units of lead have been titrated with 1 M Na_4EDTA in a continuous-flow titrator, and results agreeing within $\pm1\%$ were obtained (Priestley *et al.*, 1965).

3.17.2 *Determination of lead as oxalate*

Mayr and Fisch (1929), using the early titration method, titrated lead nitrate with oxalic acid and found that it was necessary to add 20 ml of 20% sodium acetate solution to the 150 ml of titrand to prevent interference from the nitric acid liberated in the titration. Their results had an error of $\pm0\cdot5\%$.

3.17.3 *Determination of lead as periodate*

Tejam and Haldar (1969) showed, in the titration of mercury with approximately 0.19 M $NaIO_4$, that if lead were present it could be titrated quantitatively after the mercury, in an endothermic reaction, because of the formation of the complex ion $[Pb_3(IO_6)_2]^{4-}$. Two titrations in which 43.9 and 383.8 mg of Pb^{2+} were present with mercury gave results of 43.7 and 384.6 mg respectively. The method is specific for lead, except in the presence of silver and iron which will also be titrated quantitatively. Further details of the method are given in Section 3.44.1.

3.17.4 *Determination of lead as sulphate*

Barthel *et al.* (1968) titrated 200 ml of 0.025 M $Pb(NO_3)_2$ with 0.5 M H_2SO_4 and showed that the titration curve had two not very sharp inflections. The error of the second end-point was $\pm 1\%$ and the heat of reaction obtained in the titration was $\Delta H^\circ = -1.3 \pm 0.3$ kcal/mole ($T = 0.011$ degC/mmole).

3.17.5 *Reaction of lead salts with alkalis*

(*i*) DETERMINATION OF LEAD AS CARBONATE

Priestley (1963a), using an automatic digital titrator, titrated M/60 $Pb(NO_3)_2$ with 1 M Na_2CO_3 and obtained a relative standard deviation of $\pm 1.4\%$.

(*ii*) FORMATION OF BASIC LEAD SALTS

Dutoit and Grobet (1921), using the early titration method, titrated lead nitrate with sodium hydroxide and showed that the compounds $PbO \cdot Pb(NO_3)_2$, $Pb(OH)NO_3$, $Pb(OH)_2$ and $Pb(OH)ONa$ were formed successively. The formation of $Pb(OH)_2$ was endothermic, so the inflection between it and $Pb(OH)NO_3$ was very sharp and thus it should be possible to utilize this reaction for quantitative analytical purposes.

3.17.6 *Determination of lead as sulphide*

Sajó and Sipos (1968c) in a direct-injection method for the determination of lead in alloys utilized the lead/sulphide reaction that is used primarily for the qualitative detection of lead. The sample of alloy filings (0.5 g) was dissolved in nitric and hydrochloric acids and oxidized by the addition of potassium chlorate. The solution was then neutralized with 30% sodium hydroxide solution and a known excess added to form sodium plumbite. To complex interfering ions 30% potassium cyanide solution was added and after dilution to 200 ml the solution was injected with 10 ml of sodium sulphide solution.

The titrant was a 20% solution of sodium sulphide in 10% sodium hydroxide solution. This titrant was expected to have no heat of dilution when 10 ml were injected into a blank solution but if this were not so the titrant was diluted or more sodium sulphide added according to whether the temperature rose or fell. Results obtained by the method in the analysis of alloys of known lead content are given in Table 3.22.

Table 3.22 Determination of lead in alloys
(Comparison with known contents; from Sajó and Sipos (1968c))

Number	Known content %	Found %
G Sn B2 12	0·49	0·5
LG Sn 80	2·95	3·10
GZ Rg 5	3·00	3·10
Sn 30/50	69·7	69·4
Sn 50/50	51·2	50·7
Sn 60/40	40·3	39·9

3.17.7 *Reaction of lead with fluorides*

The titration of 0·84 M NaF with 1 N $Pb(CH_3CO_2)_2$ or $Pb(NO_3)_2$ gave a titration curve with a sharp inflection showing the formation of the lead fluoride salts, which in the case of the acetate titration was attributed to the formation of PbF_2 and of the nitrate titration to $PbNO_3F$, an anologue of $PbClF$ (Deschamps *et al.*, 1968).

3.17.8 *Formation of lead(II) chromate*

The thermodynamic data for the formation of lead chromate have been determined by thermometric titration in molten salt solution (Jordan, 1960). The values obtained are summarized in Table 3.23.

Table 3.23 Thermodynamic data for the formation of lead chromate

Temperature, °C	$\Delta H°$, kcal/mole	$\Delta S°$, cal mole^{-1} degk^{-1}
168 ± 1	−18·3 ± 0·6	7·4 ± 2
250 ± 2	−20·4 ± 0·4	15·4 ± 1

3.17.9 *Formation of lead(II) nitrate and nitrite complexes*

The early titration method has been used to study the formation of lead nitrate and nitrite complexes. In the titration of alkali and ammonium nitrates with lead nitrate the compounds $4RNO_3 \cdot Pb(NO_3)_2$, $2RNO_3 \cdot Pb(NO_3)_2$ and $RNO_3 \cdot Pb(NO_3)_2$ were formed, where R was potassium, rubidium or ammonium but no complexes were formed if R were lithium or sodium (Nayar and Pande, 1951). In the titration of potassium nitrite with lead nitrate similar compounds were formed, i.e., $nKNO_2 \cdot Pb(NO_3)_2$ where $n = 4$, 2 or 1 (Vartak and Kabadi, 1955), but with sodium nitrite the compound $NaNO_2 \cdot Pb(NO_3)_2$ was formed (Vartak and Kabadi, 1954).

3.17.10 *Reaction of lead (II) with ferrocyanide*

The composition of lead (II) hexacyanoferrate(II) has been established as $Pb_2[Fe(CN)_6]$ by the thermometric titration of potassium hexacyanoferrate(II) with lead acetate, by the early titration method (Pâris and Le Chatelier, 1934; Mondain-Monval and Pâris, 1938b).

3.17.11 *Reaction of lead(II) with acetates*

The lead(II)–acetate complex has been studied, and in the titration of lead nitrate with ammonium acetate, by the early titration method, the complex ion $[PbC_2H_2O_2]^+$ was shown to be formed (Purkayastha and Sen-Sarma, 1946). For the titration of lead acetate see Section 4.1.4(*ii*).

Note

For the determination of small amounts of lead by a catalytic method see Section 3.52.

3.18 Nitrogen

The thermometric studies of inorganic nitrogen compounds discussed below include ammonia, dinitrogen tetroxide, nitrous acid, nitrates and nitric acid but not when these compounds are determined by acid or base titration, for which see Neutralization Reactions (Section 3.48).

3.18.1 *Determination of ammonia and composition of ammines*

(*i*) DETERMINATION OF AMMONIA

The exothermic oxidation of ammonia by hypochlorite has been utilized in an incremental titration method for the determination of ammonia (Brown *et al.*, 1969). It was shown that when 0.025 M NH_4^+ solution was titrated with 0.27 M OCl^-, ten replicates had a relative standard deviation of $\pm 0.09\%$.

Carr (1971a) stated that ammonium ions could be determined by titration with sodium tetraphenylborate but no results were given (see Section 3.3.2 for the conditions used).

(*ii*) COMPOSITION OF AMMINES

The composition of beryllium, copper and mercury ammine complexes has been studied by thermometric titrimetry and the results are discussed under the appropriate metal.

3.18.2 *Dinitrogen tetroxide and nitrous acid*

Using gas-injection enthalpimetry, Zambonin and Jordan (1969) have determined the heat of hydration of dinitrogen tetroxide and the heat of neutralization of the nitrous acid so formed. They also determined the heat of the Brønsted acid-base reaction of dinitrogen tetroxide and found that it was highly exothermic ($\Delta H° = -36.9 \pm 0.3$ kcal/mole). It should therefore be possible to determine dinitrogen tetroxide in this way, but carbon dioxide and sulphur dioxide (q.v.) will interfere.

3.18.3 *Nitrates*

Keily and Hume (1964) showed that lithium nitrate can be titrated in a glacial acetic acid medium with 0·5 M perchloric acid in anhydrous acetic acid. This displacement reaction was not applicable to sodium nitrate, and presumably other nitrates, because of the low solubility of these nitrates in the anhydrous medium. When a saturated aqueous solution of sodium nitrate was titrated, the reaction was endothermic and the end-point was both extremely rounded and not stoichiometric. Similar results were obtained when lithium nitrate was titrated with a partially aqueous solution of perchloric acid. Priestley (1963a) attempted to use an automatic digital titrator in the titration of ammonium nitrate with M Na_4EDTA, but there was insufficient heat from this endothermic reaction to operate the titrator.

3.18.4 *Nitric acid*

The early titration method was used by Somiya (1928b, 1932) to determine nitric acid in a mixture of water and nitric and sulphuric acids. Fuming sulphuric acid was used as the titrant and a sharp end-point was obtained corresponding to the sum of the water and nitric acid contents according to the equations $SO_3 + H_2O = H_2SO_4$ and $SO_3 + HNO_3 = SO_2(OH)(NO_3)$, respectively. A second less-distinct end-point was also obtained which represented a second reaction of the nitric acid, $3SO_3 + SO_2(OH)(NO_3) = H_2SO_4 + (SO_3)_4N_2O_5$, so that the difference in titration volumes represented the nitric acid content. The sum of the water and nitric acid contents was also determined by titration with a mixture of 510 g of 65% fuming sulphuric acid and 100 g of 61·8% nitric acid.

3.19 Phosphorus

3.19.1 *Determination of phosphorus by reaction with molybdate*

The well-known method for the determination of phosphorus involving its precipitation as ammonium phosphomolybdate in nitric acid medium has been used by Sajó and Sipos in a direct-injection thermometric method for phosphorus. However, instead of titrating the precipitate with alkali or using a reductive procedure (Vogel, 1964), they utilized the exothermic reaction of the molybdate with hydrogen peroxide as the basis of their method (see Section 3.42.1).

The precipitated phosphomolybdate was either separated, decomposed with alkali and titrated or, alternatively, it was found that the excess of molybdate added could be titrated because hydrogen peroxide had no effect on the phosphomolybdate, provided time was allowed for the precipitate to age. The former, direct, method was more suitable for contents of less than 0·1% and the latter, indirect, method for contents between 0·1 and 1·0%. The indirect method has the advantage that fewer operations are involved.

The samples (2 g) were dissolved in nitric acid to ensure that the phosphorus was present as phosphate, 3–4 ml of 3% $KMnO_4$ solution were added, the solution was boiled and the excess of permanganate destroyed with 5% $NaNO_2$ solution. A known excess of ammonium molybdate reagent was then added and the solution heated at 80 °C for a few minutes.

Direct method The solution containing the precipitate was filtered and the precipitate then washed with potassium sulphate solution until free from acid, then dissolved off the paper with 20% NaOH solution, the filtrate was collected in the original flask, acidified with hydrochloric acid, transferred to the titration vessel, diluted to 200 ml and injected with 2 ml of 30% H_2O_2.

Indirect method The solution and precipitate were allowed to stand for 5 minutes without filtration, transferred to the titration vessel, diluted to 200 ml and injected with 2 ml of 30% H_2O_2. A similar blank titration was also carried out and the sample temperature-rise subtracted from that of the blank.

The two methods were applied to the determination of the phosphorus contents of pig irons, steels and ferromanganins (Sajó and Sipos, 1967b) and of Siemens-Martin slags (Sajó and Sipos, 1966). The results given in Table 3.24 are compared with those obtained by a volumetric titration method and show good agreement even at low concentrations.

Table 3.24 Determination of % phosphorus in iron, ferromanganins and slags (Comparison of methods; from Sajó and Sipos (1966, 1967b))

Sample		Volumetric method	Enthalpimetric method	
			Direct	Indirect
Standard irons	80	0·008	0·006	—
	66	0·018	0·015	—
	30	0·021	0·025	—
	110	0·043	0·039	—
	A70	0·057	0·061	—
Pig irons	1	0·39	0·37	—
	2	0·47	0·43	—
Ferromanganins	27	0·17	—	0·15
	3	0·25	—	0·23
	9	0·47	—	0·48
	14	0·63	—	0·66
Siemens-Martin slags	1	0·35	—	0·32
	3	0·26	—	0·27
	8	0·38	—	0·37
	11	0·30	—	0·30

Note

The thermometric titration of orthophosphate and pyrophosphate with silver nitrate was not successful, because of their small heats of reaction (Harries, 1966).

The complexes of phosphoric acid with iron and those of pyrophosphate with nickel and cobalt are discussed under the appropriate sections. For the alkali titration of sodium phosphotungstates, see Section 3.45.2, and of phosphorus oxy-acids, see Section 3.48.1(*iv*). For the acid titration of alkali salts of phosphorus oxy-acids see Section 3.48.2(*iii*).

3.20 Arsenic

3.20.1 *Determination of arsenic(III) by oxidation*

(*i*) OXIDATION OF ARSENIC(III) WITH BROMATE

The volumetric determination of arsenite by oxidation with bromate has to be carried out in strongly acidic medium, but Mayr and Fisch (1929), investigating the determination as a thermometric method, using the early titration method, found, as might be expected, that there was a large heat of dilution of the hydrochloric acid solution. They found that the hydrochloric acid concentration could be reduced in the thermometric titration but not to less than 0·75%, because at lower concentrations there was a delay in the temperature rise. A hydrochloric acid range of 4–9% was found to be satisfactory, and with use of 0·1 N $KBrO_3$ titrant a solution containing 0·4948 g of As_2O_3, as determined by an iodometric method, was found to contain 0·4950 g As_2O_3.

(*ii*) OXIDATION OF ARSENIC(III) WITH IODINE

Barthel and Schmahl (1965) used a precise thermometric titration apparatus and titrated 0·025 N H_3AsO_3 in a phosphate buffer at pH 6·6 with iodine and obtained results agreeing within 0·2% and with no bias, whereas volumetrically the same titration with starch as an indicator gave results agreeing within 0·5% and with a bias of +0·2%. It is interesting to note that although this titration was successful when the phosphate buffer was used it proved to be impossible with a bicarbonate buffer; this restriction may also apply to the classical indicator titration method.

3.20.2 *Formation of aluminium arsenates*

Aluminium sulphate has been used as the titrant in an early titration method by Schroff and Kabadi (1953), who investigated the formation of gels with potassium dihydrogen, dipotassium hydrogen and tripotassium arsenate. The compounds formed were shown to be $Al_2H_3(AsO_4)_3$, $K_3Al_2(AsO_4)_2$, and $K_3Al(AsO_4)_2$ respectively.

3.21 Antimony

3.21.1 *Determination of antimony with permanganate*

Sajó and Sipos (1968a) determined antimony in its alloys with tin, lead and copper, using a direct-injection method. The sample was dissolved in sulphuric acid, and the solution diluted to 200 ml and injected with 5 ml of potassium permanganate (1 litre of saturated solution + 100 ml of water). The results obtained on samples of known antimony content are given in Table 3.25.

Table 3.25 Determination of antimony in alloys
(Comparison with known contents; from Sajó and Sipos (1968c))

Batch number	Known content %	Found %
GMS67	0·10	0·1
GZ-Rg 5	0·34	0·4
LG Pb Sn10	15·28	15·4
L Sn 30	0·14	0·2
LG Sn 80	11·43	11·7

3.21.2 *Formation of fluoride complexes of antimony*(V)

A solution of 1 N Sb^{5+} has been titrated thermometrically with 0·84 M NaF but the titration curve showed only a very curved end-point, attributed to the formation of $NaSbF_4$. The titration of the latter complex with boric acid gave a titration graph with a sharp inflection (Deschamps *et al.*, 1968).

3.22 Bismuth

The behaviour of bismuth salts in an investigation of ligand numbers is the only reference to date to the use of thermometric titrimetry in the chemistry of bismuth. This is discussed in Section 3.51.

3.23 Oxygen, water and hydrogen peroxide

3.23.1 *Determination of oxygen*

(*i*) OXYGEN IN GASES

Guérin (1952), in a paper reviewing gas analysis, described work by Ackermann in which low concentrations of between 0·1 and 1·0% of oxygen in gases were determined continuously by absorption in a circulating chromous chloride solution, using the apparatus shown in Fig. 2.18.

(*ii*) DISSOLVED OXYGEN IN HYDROCARBONS

Crompton and Cope (1968), analysing hydrocarbons for impurities that would react with triethylaluminium, showed that if oxygen were one of the impurities it could be determined down to the mg/l level. Details of the method are given in Section 3.23.2(*vi*).

3.23.2 *Determination of water*

Six methods have so far been developed for the thermometric determination of water. Of these the use of sulphuric acid as reactant is limited to a small number of materials that do not themselves react with the acid. This limitation, although not so restricting, also applies to the use of calcium hydride, triethylaluminium and to acetic anhydride; the last two reactants, for example, will react with alcohols.

The two remaining methods have a similar wide range of application both with respect to the water content and to the type of material that can be analysed. In one the heat of absorption of molecular sieves is utilized whilst in the other the well-known Karl Fischer reagent is used as reactant.

(*i*) DETERMINATION WITH SULPHURIC ACID

The highly exothermic reaction of sulphuric acid with water has been used as the basis of a number of methods for the determination of water. Somiya, using the early titration method, determined the water content of sulphuric acid (1927a, b) and of mixtures of sulphuric and nitric acids (1928b). Fuming sulphuric acid was used as the titrant and results with an error of less than $\pm0.5\%$ were obtained. A direct-injection method in which 25 ml of concentrated sulphuric acid were added to 25 g of finely powered sample contained in a Dewar has been used to determine the water content of coals, cokes, sand, ores and clays. The reaction mixture was stirred with a thermometer and the maximum temperature rise obtained was converted into water content by means of a calibration curve (Dolinskiĭ, 1937, 1938). A similar method was used by Gray and Whelan (1955), in which 5-g samples of coal were analysed. A straight-line graph of maximum temperature-rise against water content over the range 1–50% water was obtained but different rank coals had different slopes. The most accurate results were in the range 10–40% water and had a standard deviation of $\pm1.06\%$. In both these methods interference was encountered with coals of high pyrites or carbonate contents. The direct injection method was used in a reverse of the titrations described above, by Spink and Spink (1968), to determine the water content of water-miscible organic compounds such as alcohols, ketones, esters and amides. The sample, contained in a syringe, was immersed in 342 g of sulphuric acid contained in a 250-ml Dewar for about four minutes until thermal equilibrium had been established, and then 50 μl of sample were injected. It was found that 80% H_2SO_4 was the best reactant concentration as it was sufficiently active but at the same time not so viscous that it prevented effective mixing with the sample. The thermistor detected a temperature rise of approximately 0.043 degC for 100% water, and 20% v/v water in isopropanol was determined with a precision of $\pm0.3\%$. Kholler (1969) determined small quantities of water in gases by absorption in concentrated sulphuric acid in a microcalorimeter. This continuous method of measurement was sensitive to 3.4 μg of H_2O per hour.

(*ii*) DETERMINATION WITH ACETIC ANHYDRIDE

Somiya (1931, 1932) was the first to employ acetic anhydride as a reactant in the thermometric titration of water in acetic acid. Because the reaction was slow a back-titration method was employed. The sample was first reacted in a closed vessel with the anhydride for 30 minutes at 135 °C and then the excess of anhydride was titrated with aniline. Greathouse and co-workers (1956) showed that the reaction could be catalysed by the use of perchloric acid and developed both direct-injection and titration methods for water in acetic acid. In the direct-injection method, 100 ml of 2 M $(CH_3CO)_2O$ in dry acetic acid were added to 300 ml of sample and 4 ml of 2 M $HClO_4$ in dry acetic acid were then injected. Under these circumstances 3% water in the acetic acid gave a 40 degC temperature rise. In the titration method a back-titration was employed. The 2 M $HClO_4$ was added to the sample, followed by the 2 M $(CH_3CO)_2O$. The solution was then titrated incrementally with 2 M H_2O in acetic acid until there was no temperature rise. The results obtained by the two methods on known water–acetic acid mixtures and on used acetylating solutions were correct within 1% and were as precise as those obtained by the Karl Fischer method in the 2–40 g/l range of water contents. Greathouse (1957) used the methods to determine the water content of vegetables by extracting them with dry acetic acid. A similar method has been used by Russian workers to determine water in organic solvents (Goizman *et al.*, 1971).

(*iii*) DETERMINATION WITH CALCIUM HYDRIDE

The three calcium salts, chloride, carbide and hydride, combine exothermally with water, the heats of reaction being $-3\cdot65$, -15 and $-27\cdot5$ kcal/mole H_2O respectively. Calcium hydride was therefore chosen as the reactant in a thermometric method for water in petroleum products (Orzherovskii, 1969). A test-tube, 135 mm long and 30 mm in diameter, with an insulated bottom, and marked at the 10-ml level, was used as the reaction vessel. The sample, either 10 g or 10 ml, was placed in the tube and after a few minutes the temperature was recorded. Up to 0·7 g of finely powdered calcium hydride, weighed and sealed into a small glass ampoule, was added to the sample. The thermometer was used to break the ampoule and mix the calcium hydride with the sample. The maximum temperature was recorded and the water content read from a calibration curve. The method, which was applicable to water contents over the range 0·05–2·0%, was shown to determine the known water content of 24 turbine and motor oils to within $\pm0\cdot05\%$. The absence of bubbles of hydrogen escaping from the sample was stated to show that the sample contained less than 0·05% of water.

(*iv*) DETERMINATION WITH TRIETHYLALUMINIUM

Crompton and Cope (1968), using an intermittent thermometric titration, determined the total reactants in hydrocarbons with triethylaluminium solution in the hydrocarbon (80 g/l) as titrant. In the work, ethyl alcohol was the reactant of main interest and was used to calibrate the apparatus, but water and oxygen also reacted with equal molar sensitivity to that of alcohol and could therefore also be determined down to the mg/l level.

(*v*) DETERMINATION WITH KARL FISCHER REAGENT

The Karl Fischer method is probably the one most frequently used for the determination of water, but in spite of the large exothermic reaction of the reagent with water (-16 kcal/mole H_2O) it was not until 1966 that Wasilewski and Miller utilized the reaction in a direct-injection method.

The method consisted of three stages, in each of which 20 μl of water (a fourfold excess) were injected into 1·0 ml of Karl Fischer reagent which had a titre of about 5 μl of water per ml. In the first stage 1·0 ml of dried sample was added (ΔT_o), in the second stage 1·0 ml of sample was added (ΔT_s), and in the third stage 1·0 ml of a solution of known water content was added (ΔT_w). The water content was calculated from the temperature rises (ΔT) by the equation in which W_w is the water content of the known solution.

$$W = W_w . \frac{\Delta T_o - \Delta T_s}{\Delta T_o - \Delta T_w}$$

The authors state that the features of this method are that the water content is correlated directly with a definite electrical signal and the measurement is made in the presence of excess of water, enabling the reaction to go rapidly to virtual completion. They claim the method to be applicable to all types of material because gases can be injected into the Karl Fischer reagent and the heat pulse measured; solids can be equilibrated with reagents before the injection of the water.

The method, however, is very dependent on there being high precision in the injection of the 20 μl of water in each stage and this was achieved with an air-operated burette (Marinenko and Taylor, 1966). Water contents of between 0·030% and 0·234% were determined with a precision of better than $\pm 2\%$ and a relative error of less than $\pm 4\%$. This error should be compared with that of $\pm 2·8\%$ obtained by Bonner (1946) when water contents of 0·31–0·06% in dioxan were titrated volumetrically with Karl Fischer reagent. Large errors were obtained in a direct-injection method (Erdey and Marik, 1970).

(*vi*) DETERMINATION WITH MOLECULAR SIEVES

The absorption of water by molecular sieves is exothermic and this has been used by Reynolds and Harris (1969) in a direct-injection method for water in organic liquids. The liquid being examined (25 ml) and molecular sieve Linde Type 4A (0·6 g) were placed in a stirred insulated cell, and a 1·0 ml hypodermic syringe was used to inject 0·2 ml of the sample into it. The cell was calibrated by injection of the same liquid containing known quantities of water; solids were examined by equilibrating them in dimethylformamide. It was found that acetonitrile and methanol could not be examined as they were absorbed by the sieves because of their small size, and that glycerol was too viscous to be stirred effectively.

3.23.3 *Determination of hydrogen peroxide*

(*i*) DETERMINATION WITH PERMANGANATE

The early titration method has been used to determine hydrogen peroxide with permanganate under conditions similar to the well-known volumetric method (Mayr and Fisch, 1929). Between 20 and 40 ml of 0·1 N H_2O_2 were taken, 10 ml of dilute (1 + 4) sulphuric acid were added, and the whole was diluted to 60 ml and titrated with 0·1 N $KMnO_4$. No results were given in the paper.

(*ii*) DETERMINATION WITH CERIUM(IV)

The results obtained in the titration of 0·008 N H_2O_2 with 0·35 N Ce(IV) in 3 M H_2SO_4, using a differential thermometric apparatus, were not very satisfactory. Errors of $+3·7\%$ were obtained which were attributed to the slow speed of the reaction, which extended the titration curve beyond the true equivalence point (Tyson *et al.*, 1961). It should be feasible to utilize this reaction by using a suitable catalyst.

3.24 Sulphur

3.24.1 *Determination of sulphur by oxidation of hydrogen sulphide*

The oxidation of sulphide ions to sulphate ions by potassium permanganate in alkaline solution is an extremely exothermic reaction ($\Delta H = -100$ kcal/mole, i.e., over six times greater than a neutralization reaction) and consequently extremely sensitive thermometric methods for sulphide can be based on it. If sulphur in a substance can be converted into hydrogen sulphide the reaction provides a sensitive method for the sulphur, and Sajó and Sipos (1966, 1967b) utilized this reaction to determine sulphur in iron, steels and slags by a direct-injection method and stated that contents as low as 0·005% could be determined with a 5-g sample.

The samples, 5 g or 1 g according to the content, were dispersed in water and hydrochloric acid was added. The liberated hydrogen sulphide was absorbed in 100 ml of 20% NaOH solution and nitrogen was passed through the reaction mixture at the end of the reaction to remove the last traces of hydrogen sulphide. The sulphide solution was diluted to 200 ml and 5 ml of saturated potassium permanganate were injected into it.

The method was compared with an iodometric method for the sulphide content, and the results given in Table 3.26 show that good agreement was obtained between the methods. The thermometric method has also proved to be superior to the volumetric titration of hydrogen sulphide with permanganate, which is decidedly unsatisfactory (Bethge, 1953).

Table 3.26 Determination of sulphur in pig-irons, steels and slags
(Comparison of methods; from Sajó and Sipos (1966, 1967b))

Material		*Iodometric method, %*	*Enthalpimetric method, %*
Pig irons	NA	0·027	0·024
	NL	0·056	0·060
	GG1	0·094	0·098
Steels	S50	0·007	0·006
	S66	0·013	0·011
Blast-furnace slags	1	0·89	0·83
	2	1·22	1·24
	3	1·23	1·30
	4	1·02	0·97
	5	1·17	1·10
	6	0·74	0·81

3.24.2 *Determination of hydrogen sulphide*

Guérin (1952) reported that hydrogen sulphide in gas streams could be determined continuously by absorption in a circulating sodium hypochlorite solution, the temperature of which was measured (see Fig. 2.18).

3.24.3 *Determination of sulphur dioxide*

Zambonin and Jordan (1969), investigating gas-injection enthalpimetry, showed that the injection of sulphur dioxide into 0·02–0·1 M KOH was ideally suited for the quantitative analysis of gases for their sulphur dioxide content; the equilibrium was virtually complete, the kinetics were fast and the heat of reaction ($\Delta H^\circ = -39$ kcal/mole) high enough for the method to be sensitive. On a small scale as little as 0·5 μmole of SO_2 could be determined, but the lowest concentration that could be determined with an error of the order of 2% was 1% SO_2. The sample had to have the same water-vapour pressure as that of the potassium hydroxide solution, otherwise a blank injection had to be made of a neutral gas with the same water-vapour pressure as that of the sample. Carbon dioxide and nitric oxide will obviously interfere.

3.24.4 *Determination of sulphites*

Oxidation is a common means for the determination of sulphites; three oxidants have been investigated for this purpose. These are hydrogen peroxide, cerium(IV) and permanganate, but iodine, as in volumetric analysis, has only been used in a back-titration procedure in which excess of iodine is titrated with thiosulphate, as described in Section 3.29.1.

(*i*) DETERMINATION BY OXIDATION WITH HYDROGEN PEROXIDE

Concentrated ammonium sulphite and bisulphite (hydrogen sulphite) solutions obtained during industrial absorption of sulphur dioxide in concentrated ammonium sulphite solution have been analysed, using a continuous titrator, by oxidation with hydrogen peroxide (Stráfelda and Hájková, 1968). Tests with sulphite solutions with concentration ranges near to 0·4 N and to 7·0 N gave linear calibration graphs in both cases, showing that the effect of heat of dilution with the 30% H_2O_2 titrant was negligible. The results were reproducible to within $\pm$0·65%.

(*ii*) DETERMINATION BY OXIDATION WITH CERIUM(IV)

Priestley (1963a), using an automatic digital titrator, titrated M/240 Na_2SO_3 with M/4 cerium sulphate in sulphuric acid and in thirteen replicate titrations obtained a standard deviation of $\pm$0·7%.

(*iii*) DETERMINATION BY OXIDATION WITH PERMANGANATE

In another automatic titration method 0·2 mmole of Na_2SO_3 in 20 ml of solution were titrated with 0·05 M $KMnO_4$, and as in the permanganate oxidation of sulphide above the heat of reaction was found to be six times that of a neutralization reaction, and consequently a large T value of 0·37 degC/mmole was obtained in the titration (Priestley *et al.*, 1963).

3.24.5 *Determination of sulphate as barium sulphate*

(*i*) THERMOMETRIC METHODS

The classical method for the determination of sulphate by precipitation as barium sulphate has also been investigated as a thermometric method by a number of workers. Dean and Watts (1924) were the first; using the early titration method they titrated solutions of sulphate with 0·5 N BaCl$_2$ and obtained results correct to within 5%. Results obtained on two solutions and on a sample of chalcopyrite of known sulphur content are given in Table 3.27. The pyrite ore was fused with sodium peroxide to obtain the solution for analysis. Later work by Mayr and Fisch (1929) showed that the precision of the method was affected by the cation present; the presence of sodium ions gave good results, but potassium ions, particularly at high concentration, gave poor ones. The method was not applicable in the presence of magnesium ions. This was attributed to the adsorption of the soluble sulphate on the precipitate.

Recent work by Williams and Janata (1970) has shown that the stoichiometry of the precipitation from solutions containing sodium ions can be improved even further by preventing the adsorption of soluble unprecipitated sulphate on the freshly precipitated barium sulphate by the use of alcoholic solutions. Solutions containing above 50% alcohol were found to give the best results, and 70% solutions were used for a series of tests in which organic sulphur-containing samples were first decomposed by sodium peroxide fusion or flask combustion.

No interference was encountered from iron, copper, nickel or silver. Interference from fluoride was removed by the addition of boric acid. The results obtained are given in Table 3.27, and it would appear that some error is being introduced in the flask combustion procedure.

Table 3.27 Determination of sulphur by thermometric titration

Sample	*Quantity* mg	*Sulphur content, %*		*Relative standard deviation* ±%	*Reference*
		Known	*Found*		
Sulphate solution	500	20·49	20·32	—	(*a*)
Sulphate solution	500	14·11	14·14	—	(*a*)
Pyrite ore	500	43·92	43·70	—	(*a*)
Sulphuric acid	0·981	32·7	31·8	1·6	(*b*)
Sulphuric acid	9·804	32·7	32·7	1·1	(*b*)
Sodium sulphate	14·20	22·5	22·9	1·8	(*b*)
Sulphonal	10–13	28·1	28·9*	7·1	(*b*)
Sulphonal	10–13	28·1	27·0†	1·6	(*b*)
Benzyl thiouronium chloride	19–24	15·8	16·3*	2·7	(*b*)

* Preceded by a flask combustion
† Sodium peroxide fusion
(*a*) Dean and Watts (1924)   (*b*) Williams and Janata (1970)

Further studies of this reaction have been made by Perchec and Gilot (1964) and by Barthel and co-workers (1968). The former used a thermo-pile/piston-burette apparatus and titrated 100 ml of acidified 0·01 M sulphate with either 0·2 M $Ba(NO_3)_2$ or 1·0 M $BaCl_2$ and obtained a precision of $\pm0\cdot5\%$. The acidity of the solution was not important except that it had an effect on the heat of dilution. They also investigated the use of derivative titration curves (according to Zenchelsky and Segatto, 1957b), which gave a very sharp end-point enabling solutions containing 0·01 mole of SO_4^{2-} to be analysed. The latter authors used a precise thermometric titration apparatus and with barium nitrate titrated 0·05 M Na_2SO_4 or 0·015 M H_2SO_4 and obtained precisions of $\pm0\cdot2\%$ and $\pm0\cdot3\%$ and enthalpy changes $(\Delta H°)$ of $-4\cdot55 \pm 0\cdot1$ and $-7\cdot78 \pm 0\cdot02$ kcal/mole respectively.

(ii) ENTHALPIMETRIC METHODS

Sajó and Sipos have used a direct-injection method to determine the sulphate content of plating-bath solutions (1967a) and of cements and clinkers (1968a, 1969b). In order to obtain a rapid temperature rise, a neutral suspension of barium sulphate containing 25–30 g of $BaSO_4$ per 100 ml was added to the test solutions to provide nuclei for the precipi-tation. The barium chloride titrant (39·5 g of $BaCl_2.2H_2O$ + 30·5 ml of conc. hydrochloric acid + 155 ml of water) was expected to have no

Table 3.28 Determination of sulphate content of plating solutions, clinkers and cements

(Comparison of methods; from Sajó and Sipos (1967a, 1968a))

Sample		Gravimetric method, g/l	Enthalpimetric method, g/l
Nickel plating solutions	1	71·3	73·0
	2	91·5	92·0
	3	164·7	165·0
	4	109·8	109·0
	5	126·1	124·0
	6	146·4	145·0
Chromium plating solutions	1	435·2	434·0
	2	397·4	399·0
	3	371·3	321·0
	4	169·2	168·0
	5	157·3	158·0
	6	158·1	158·0
Clinkers*	1	0·22	0·30
	2	0·18	0·20
Cements*	1	3·12	3·0
	2	3·50	3·3

* Results calculated as % SO_3

heat of dilution when injected into a blank solution. If the solution temperature fell or rose either hydrochloric acid or barium chloride was added until there was no heat of dilution. To analyse plating solutions, 5 ml of the solution were taken, 20 ml of hydrochloric acid and 5 ml of the neutral barium sulphate suspension were added and the whole diluted to 200 ml; 10 ml of titrant were then injected. It was found that chromic acid interfered and if this were present a correction of 0.05% SO_4^{2-}/g H_2CrO_4 had to be added to the determined content. To analyse clinkers and cements 2 g of the sample were dissolved in 60 ml of hydrochloric acid and the solution was diluted to 200 ml; then 5 ml of titrant were injected. The results obtained, compared with those from the classical gravimetric method, are given in Table 3.28 and show good agreement between the methods.

(*iii*) DUAL TITRATION METHOD

Taubinger (1969b) has used the dual titration method to determine sulphate ions in the presence of vanadium(V) and titanium(IV) ions by titration with sulphuric acid and using barium chloride solution containing a suspension of barium sulphate as the reactant.

3.24.6 *Determination of sulphuric acid*

The determination of sulphuric acid is commonly carried out by titration with alkali and this is discussed in Section 3.48, but other methods for its determination at high concentrations by reaction with water and with barium acetate are discussed below.

(*i*) DETERMINATION BY REACTION WITH WATER

In what is probably the eariliest reference to an enthalpimetric method Howard (1910) described the determination of the strengths of fuming sulphuric acid. The sample was reacted with a solution of water in sulphuric acid (this solution approximated to sulphuric acid monohydrate), the total analysis time taking less than four minutes. For example, 100 g of sample containing $21–26\%$ free sulphur trioxide were placed in a Dewar flask and 100 g of 92% sulphuric acid added to it. Under these conditions T was 22–39 degC. Conversely for the analysis of the monohydrate, fuming sulphuric acid was injected and over the range $94–99\%$ H_2SO_4 T was 2–10 degC.

Richmond and Merryweather (1917) used a different procedure to assay sulphuric acids in that they injected 5 ml of sample into a Dewar flask containing 400 ml of water and detected the temperature rise with a thermometer graduated in 0.01 degC. The results on samples containing $92.4–97.4\%$ H_2SO_4 were correct within $+0.1\%$ absolute, when compared with a volumetric method.

(*ii*) DETERMINATION BY REACTION WITH BARIUM ACETATE

Somiya (1928a) described the use of the early titration method in the determination of sulphuric acid in admixture with acetic acid and acetic anhydride, i.e., in cellulose acetylation solutions, by titration with barium acetate in solution in acetic acid containing about 5% acetic anhydride. This non-aqueous titration was standardized with solutions of known concentrations of sulphuric acid in acetic acid.

3.24.7 *Reactions of sulphates and persulphates*

(*i*) INTERACTION OF SULPHATE IONS WITH OTHER IONS

The interaction of sulphate ions with H^+, Na^+ and K^+ ions and with 26 bi- and tervalent ions has been determined by titration of 0·02 M metal chlorate solutions with tetra-alkyl ammonium sulphate solutions (alkyl = Me, Et, n-Pr) and the log K, $\Delta H°$ and $\Delta S°$ values for the interaction have been evaluated at 25 °C (Izatt *et al.*, 1969). The results indicated that the quaternary salts did not associate to any appreciable extent with the sulphate ion.

(*ii*) ACTION OF BORON TRIFLUORIDE ON SULPHATE AND PERSULPHATE

Richards and Woolf (1969) have studied the reaction of boron trifluoride containing bromine on potassium sulphate and persulphate. In both cases oxygen and potassium fluorosulphate were formed and, in the case of potassium sulphate, potassium fluoride was also formed.

(*iii*) DETERMINATION OF PERSULPHATE BY REDUCTION

The highly exothermic ($\Delta H° = -34 \pm 2$ kcal/mole) reduction of per-sulphate by ferrous ions has been used as a means of determining per-sulphate, using a precise thermometric titration apparatus (Barthel and Schmahl, 1965). They titrated 0·0125 N $(NH_4)_2S_2O_8$ in 0·1–1·0 N H_2SO_4 solution with 0·25 M $(NH_4)_2Fe(SO_4)_2$ and obtained results accurate within $\pm0.05\%$. A similar precision was obtained in a volumetric back-titration method using permanganate, but the results had a bias of -0.2%.

3.24.8 *Determination and composition of thiosulphates*

(*i*) DETERMINATION OF THIOSULPHATE BY HYPOCHLORITE OXIDATION

Using a high-precision incremental titration procedure, 0·0125 M $S_2O_3{}^{2-}$ has been titrated with 0·27 M OCl, and in five replicate titrations a relative standard deviation of $\pm0.06\%$ was obtained (Brown *et al.*, 1969).

(*ii*) DETERMINATION OF THIOSULPHATE BY IODINE OXIDATION

Barthel and Schmahl (1965), using a precise thermometric titrator, titrated 0·01 N $Na_2S_2O_3$ with 0·2 N I_2 and obtained results precise to within $\pm0.2\%$, compared with $\pm0.3\%$ by a volumetric starch-indicator method; both methods were accurate. Priestley and his co-workers (1968) reported that they had titrated 0·5 M $Na_2S_2O_3$ with KI_3, using a continuous-flow titrator.

(*iii*) DETERMINATION OF THIOSULPHATE WITH ACID

Priestley (1963a) used an automatic digital titrator to titrate $M/60$ $Na_2S_2O_3$ with 1 M HCl and in five replicate titrations obtained a relative standard deviation of $\pm 0.8\%$.

(*iv*) COMPOSITION OF THIOSULPHATES

The composition of cadmium, copper(II) and silver thiosulphates has been determined and is discussed under the appropriate cation.

3.24.9 *Formation of chloro- and fluorosulphuric acids*

The heats of formation of chloro- and fluorosulphuric acids were determined by reaction of sulphur trioxide with excess of hydrogen chloride or fluoride dissolved in fluorosulphuric acid and found to be -142.9 and -189.4 kcal/mole respectively (Richards and Woolf, 1967). The same authors (1968) also determined the heat of formation of the potassium, sodium and copper(II) salts of nitrosonium- and nitronium-fluorosulphates.

3.25 Selenium

3.25.1 *Determination of selenium*

In an investigation into the thermometric titration of selenium as selenite, Dastoor and Haldar (1967) studied the use of silver ions, lead ions, bismuth ions and oxidizing agents as titrants; their use of alkali as titrant is discussed in Neutralization Reactions, Section 3.48.1(*vii*).

(*i*) TITRATION OF SELENITES WITH HEAVY METAL IONS

In the titration of alkali selenites or selenious acid with solutions of the nitrates of bismuth or silver it was found that the bismuth reaction had too small a heat of reaction. With silver and lead nitrates sharp inflections were obtained, equivalent to the ratios 2Ag : 1Se and 1Pb : 1Se respectively at the end-points. It was shown that by using either silver or lead nitrate as titrant 33 mg of Se in 30 ml of solution could be determined with an error of $\pm 0.5\%$.

(*ii*) TITRATION OF SELENIOUS ACID WITH OXIDIZING AGENTS

Both potassium persulphate and sodium hypochlorite were found to be unsatisfactory as titrants; the former gave a very poor end-point and the latter had too small a heat of reaction. With potassium permanganate as titrant two end-points were found; the first was equivalent to an H_2SeO_3 : $KMnO_4$ ratio of 5 : 2 but the second was dependent on the solution concentrations. The first inflection was used in the determination of selenium in the presence of other species, with the results shown in Table 3.29.

Interference with the method by the presence of the reducing ions Fe_2^+, As^{3+}, Sb^{3+}, TeO_3^{2-}, SO_3^{2-}, S^{2-} and $S_2O_3^{2-}$ can be removed by a prior oxidation with potassium dichromate.

It follows that ternary mixtures of selenious and tellurous acid with sulphuric acid, nitric acid or selenic acid could be analysed by three successive thermometric titrations: (i) for the total acidity by alkali titration

Table 3.29 Determination of selenium in the presence of other species
(From Dastoor and Haldar (1967))

| Species present | | Selenium mg/30 ml | |
		Present	Found
—		20·7	20·85
—		165·6	166·6
H_2SO_4	0·5 N	57·0	57·2
HNO_3	0·5 N	41·4	41·1
NO_3^-	318 mg	123·9	124·1
Zn^{2+}	170 mg	103·3	102·9
Cd^{2+}	120 mg	90·9	91·5
Cu^{2+}	63 mg	61·7	61·6
TeO_4^{2-}	190 mg	38·8	38·6

(c.f. Section 3.48.1(vii)), (ii) for the tellurous acid content by dichromate titration, and (iii) for the selenium content by permanganate titration.

3.25.2 *Determination of selenium(IV) with* QDT

A direct-injection method for the determination of selenium (IV) with quinoxaline-2,3-dithiol (H_2DQT) as reactant has been developed by Beezer and Slawinski (1971). When 0·6 ml of 1 M $Na_2QDT.2H_2O$ was injected into 25 ml of selenium(IV) solution in 0·03 M $HClO_4$ the results in Table 3.30 were obtained.

Table 3.30 Precision of results in the determination of selenium(IV)

Concentration	No. of replicates	Standard deviation ±%
M/167 Se(IV)	5	0·3
M/415 Se(IV)	5	0·6

The method was stated to be superior to the absorptiometric method using the same reagent.

3.26 Fluorine

3.26.1 *Determination of fluoride—thermometric methods*

Priestley and his co-workers (1963) showed that fluoride could be titrated with thorium nitrate in an automatic apparatus. If 20 ml of a solution containing 1 mmole each of the four halides were titrated with a mixed titrant [1 M $Th(NO_3)_4$ + $AgNO_3$] the fluoride was titrated first ($T = 0.4$ degC/mmole) followed by each of the other halides (see Fig. 3.3). In a later paper, Everson and Ramirez (1967) showed that, as well as thorium, a number of other ions were suitable as titrants for fluoride; these included cerium(III), aluminium and calcium ions. A differential titration technique was used to minimize heats of dilution, and it was shown that by a suitable selection of titrant and pH of solution it was possible to titrate fluoride in the presence of moderate amounts of sulphate, borate, phosphate or silicate. Thus in many cases the time-consuming Willard–Winter distillation purification procedure was avoided, but the authors recommended a zinc acetate pretreatment to avoid interference from silicate, phosphate and iron. The results, obtained by using 0.2–1.0 M Ca^{2+} as titrant, given in Table 3.31 were obtained by adding 5% zinc acetate solution to the sample, making it just acid to phenolphthalein, adding 3 ml of 10% ammonium acetate solution, diluting to 25–30 ml and finally diluting with an equal volume of 50% aqueous propan-2-ol. The reference solution contained 20 ml of water, 25 ml of propan-2-ol and 5 ml of triethanolamine.

Table 3.31 Determination of fluoride in the presence of interfering ions
(Selected results; from Everson and Ramirez (1967))

Fluoride added, mg	Interfering ions added, mg		Zinc acetate soln. added, ml	Fluoride found, mg
24·9	none	—	2–10	24·3–25·1
24·9	$HSiO_3^-$	25	4	25·5
24·9	$HSiO_3^-$	100	8	25·8
24·9	HPO_4^{2-}	42	3	24·7
24·9	HPO_4^{2-}	85	6	24·4
24·9	Fe^{3+}	35	5	25·8
24·9	$\begin{cases} HPO_4^{2-} & 38 \\ HSiO_3^- & 50 \end{cases}$		8	25·2
10·0	$HSiO_3^-$	25	4	10·2
10·0	$HSiO_3^-$	100	6	11·0
10·0	HPO_4^{2-}	42	3	10·1
10·0	HPO_4^{2-}	85	5	10·0

Aluminium interferes with the calcium titration method but this can be avoided by using aluminium ions as the titrant.

Everson and Ramirez also found that their thermometric method was useful because they could investigate the course and nature of the metal–fluoride reaction. For example, the titration curve obtained when Al^{3+} ions were titrated with fluoride in aqueous propan-2-ol solution showed four inflections caused by the formation of AlF_4^-, an unknown, AlF_5^{2-} and AlF_6^{3-} ions, the first and last inflections being quite sharp.

3.26.2 *Determination of fluoride—enthalpimetric and continuous-flow methods*

C. E. Johansson (1970) has utilized the fast exothermic precipitation reaction $Pb^{2+} + F^- + Cl^- \rightarrow PbFCl$ in a direct-injection method. The carbonate-free sample was passed through an anion- and cation-exchange resin system, the resultant hydrofluoric acid being absorbed in a sodium acetate–hydroxide–chloride solution. This solution was evaporated to 6–7 ml, brought to pH 4·4, diluted to 25 ml and injected with 0·83 ml of 4·9 M $Pb(ClO_4)_2$. Results on samples containing 6–8 mg of F^- were correct within $\pm2\%$, an error similar to that found in the semi-micro gravimetric method of Belcher and Tatlow (1951). The authors also determined the ΔH value for the precipitation of PbFCl from water solutions with an ionic strength of 1 M as $36·6 \pm 0·3$ kJ/mole.

Similarly prepared solutions but containing between 40 and 400 μg of F^- were analysed, an LKB commercial microflow calorimeter and 0·49 M $Pb(ClO_4)_2$ as titrant being used. The results obtained had an error of ±4 μg F^-. Concentrations higher than 4 mM could not be analysed because the precipitate caused frictional heating in the calorimeter or stopped the flow.

Note

The thermometric studies of the fluorine compounds boron trifluoride, fluorosulphuric acids and fluoroberyllates are discussed in the appropriate sections.

3.27 Chlorine

3.27.1 *Determination of chloride*

(*i*) As SILVER CHLORIDE

The precipitation of silver chloride, which is the basis of the classical gravimetric and volumetric methods for chlorides, has also been extensively studied as a basis for the thermometric method, both in the titration with silver, discussed below, and with chlorides discussed under Silver in Section 3.39.1. Workers using the early titration method found that sharp end-points could be obtained with 0·2–0·5 M $AgNO_3$ as titrant. Dean and Newcomer (1925) obtained results within $\pm1·5\%$ of the gravimetric results and showed that cyanides, bromides and iodides could also be titrated with the same precision but could not be distinguished from the chloride or each other. Mayr and Fisch (1929) titrated sodium, potassium, ammonium and calcium chlorides and reported that the cation present had an effect on the precision of the titration of the chloride, potassium giving very variable results. Priestley and co-workers (1968) titrated sodium chloride solutions with 0·2 M and 1 M $AgNO_3$, using a continuous-flow titrator. Chlorides have been successfully titrated in a burette–thermopile apparatus (Harries, 1966). A relative standard deviation of $\pm0·64\%$ was obtained when 0·01 M KCl was titrated with 1 M $AgNO_3$ but at lower concentrations the precision was considerably poorer (Harries, quoted in Tyrrell and Beezer, 1968). Very precise results were obtained by an incremental titration procedure (Brown *et al.*, 1969). A relative standard deviation of $\pm 0.019\%$ was obtained in ten replicate titrations of 0·038 M Cl^- with 0·5 M Ag^+.

The interference of bromide and iodide in the titration of chloride is important and the possibility of this occurring can be judged from the solubility products and heats of precipitation of the silver halides given in Table 3.32 in which the results obtained by Ewing and Mazac (1966) using thermometric titrimetry are shown to compare very favourably with the literature values.

Table 3.32 Properties of silver halides

	Solubility product	Heat of precipitation kcal/mole at 25 °C	
		Ewing and Mazac	Literature
Silver iodide	$1·5 \times 10^{-15}$	$-26·94$	$-26·85$
Silver bromide	$7·7 \times 10^{-13}$	$-19·90$	$-20·19$
Silver chloride	$1·6 \times 10^{-10}$	$-15·70$	$-15·65$

These figures show that the three halides should be distinguishable in a thermometric titration giving sufficient difference in the slopes of the

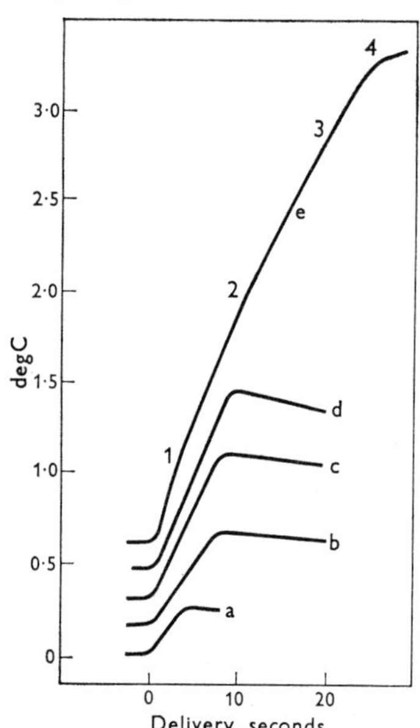

FIG. 3.3 Titration of halides with M $Th(NO_3)_4 + AgNO_3$ in an automatic titrator; from Priestley *et al.* (1963).
a NaF; b NaCl; c NaBr; d NaI; e $NaF + NaI + NaBr + NaCl + HNO_3$; 1 mmole in 20 ml.

titration graphs of each halide. That this separation is possible was shown, using an automatic thermometric titration (Priestley *et al.*, 1963) in which all the halides were titrated singly or in admixture in 20 ml of nitric acid solution containing 1 mmole of the halide, with M $Th(NO_3)_3 + AgNO_3$ as titrant. The titration graphs obtained are given in Fig. 3.3, which shows the difference in slopes of silver precipitation. As would be expected from the solubility products the titration proceeds in the order iodide, bromide and chloride. The lack of sharpness at the iodide and bromide end-points implies that small quantities of one halide cannot be distinguished in the presence of the others but that reasonable titrations of chloride and iodide should be possible in admixture of the two.

The work of Harries (1966) showed, however, that even with the use of a precision apparatus the results of the titration of mixtures of these halides were inaccurate. This was attributed to co-precipitation (small in an iodide–bromide titration and large in a bromide–chloride titration). The amount of co-precipitation was dependent on the ion ratio. For example, the error was $\pm 1\%$ with a $1:1$ ion ratio but $\pm 10\%$ with a $4:1$ ion ratio in an iodide–bromide titration; in a chloride–bromide titration with an ion ratio of $1:4$ the error was as big as almost 40% in the chloride determination. The determination of the total halide content in all cases of mixed halide titrations was found to have an error of less than $\pm 1\%$. Recently Takeuchi and Yamazaki (1969a), using an apparatus employing a different thermistor system, showed that an error and precision of 1–3% could be obtained in the titration of 0.01 mole/l of each component in binary mixtures (chloride–iodide, bromide–iodide and chloride–bromide) and also in a ternary mixture of all three ions. Similar precision was also obtained in the titration of a binary mixture of iodide and thiocyanate.

High-temperature titration of between 8×10^{-4} and 2×10^{-2} molal chloride with 1–1.5 molal silver nitrate has been carried out in molten lithium nitrate–potassium nitrate ($34:66$ w/w). At 158 ± 2 °C the results agreed within $\pm 4\%$ and the heat of precipitation was estimated as -18.9 kcal/mole (Jordan *et al.*, 1959a).

Sajó and Sipos (1967a) determined the chloride content of nickel plating solutions, using a direct-injection method. To 15 ml of the plating solution were added 15 ml of concentrated nitric acid, the solution was diluted to 200 ml and 5 ml of silver nitrate titrant solution were injected. The titrant, prepared by dissolving 200 g of silver nitrate in 400 ml of water and adding 96 ml of ethanol and 50 ml of concentrated nitric acid was expected to show no heat of dilution when it was injected into a blank solution. If the temperature rose, more silver nitrate was added to the titrant, and if it fell, more alcohol. From 2 to 5 g/l of Cl^- in plating solutions could be determined to within ± 0.10 g/l.

Ewing and Mazac (1966) showed that the heats of precipitation of the silver halides are dependent on the ionic concentration in the solution from which the precipitation takes place. In the ionic concentration range from infinite dilution to 0.01 M the heat of precipitation of AgCl changed by 0.4 kcal/mole (2.5%), that of AgBr by 1.2 kcal/mole (6%), and that of AgI by 1.0 kcal/mole (4%). These changes can only be caused by adsorption effects and this probably accounts for the poor precision of the methods generally, and in particular of the direct-injection method where adsorption effects are much more likely. In the Ewing and Mazac experiments silver was titrated with halide and the general precision of these titrations is better than the reverse titration discussed here (see Table 3.59) from which it can be concluded that adsorption effects are more marked in the titration of halides with silver.

(ii) AS PERCHLORATE

Keily and Hume (1964) showed that if a slight excess of mercury(II) acetate were added to a solution of lithium chloride in acetic acid, lithium acetate was formed which could then be titrated with 0·5 M perchloric acid in solution in anhydrous acetic acid or in acetic acid sufficiently aqueous to be equimolar in both perchloric acid and water (hydronium perchlorate). With the latter titrant a relative standard deviation of $\pm0\cdot24\%$ was obtained for the precision and the error for both titrants was $\pm0\cdot3\%$. Very similar results were obtained in the titration of sodium chloride with the 'hydronium perchlorate' titrant (see also Section 4.1.4(ii)).

(iii) AS MERCUROUS CHLORIDE

Harries (1966) successfully titrated chlorides with mercurous nitrate solution, using a burette–thermopile apparatus.

3.27.2 *Determination of hypochlorite by reduction*

Reducing agents have been used as titrants in the thermometric determination of hypochlorite. Mayr and Fisch (1929), using the early titration method, found that arsenious acid as titrant gave sharp and reproducible end-points. Recently Štráfelda and Kroftová (1968b), investigating the determination of the hypochlorite content of fabric-bleaching solutions, found that 10% sodium sulphide or 5% sodium thiosulphate solutions were satisfactory titrants for use with a continuous titrator. The heats of dilution of the titrants in the bleaching-bath solutions were shown to be about 1% of the heat of reaction being measured. The sensitivity with the sodium sulphide titrant was slightly higher, but with both titrants it was possible to determine the lowest expected concentration of active chlorine. The temperature rise plotted against the active chlorine content in the range 0–4 g was linear for mixing rates of titrant and bath solution of between 1·25 and 13·35 ml/min except at the lowest rate and smallest content; this was attributed to poor mixing.

A direct-injection method has been developed for the determination of chlorine in water, using the Directhermom apparatus. The determination was made either directly by the injection, into 20 ml of sample, of 2 ml of potassium iodide solution which had no heat of dilution (10 ml of concentrated HCl + 90 ml of 20% KI solution) or indirectly but with less sensitivity by the injection (after the addition of the iodide) of 2 ml of sodium thiosulphate solution which had no heat of dilution (65 ml of N $Na_2S_2O_3$ + 35 ml of glycerol). The method was applied to the analysis of bleaching powder, and contents of 24–30% of available chlorine were determined to within $\pm 1\%$ absolute (Marik-Korda and Erdey, 1970).

Note

The titration of hydrochloric and chlorine oxy-acids with bases is discussed under Neutralization Reactions in Section 3.48.1(*iii*). For the determination of small amounts of chloride by a catalytic method see Section 3.52.

3.28 Bromine

3.28.1 *Determination of bromide*

(*i*) AS SILVER BROMIDE

Although it has been shown that bromide can be satisfactorily titrated with silver nitrate (Harries, 1966) its titration in the presence of other halides is difficult because the end-points on the titration graphs between it and chloride or iodide are not very sharp. This is discussed more fully in Section 3.27.1.

(*ii*) AS MERCUROUS BROMIDE

Harries (1966) successfully titrated bromides with mercurous nitrate solution, using a burette–thermopile apparatus.

3.28.2 *Determination of hypobromite by reduction*

Mayr and Fisch (1929), using the early titration method, showed that arsenious acid could be used as a reducing titrant and obtained a sharp end-point in the titration of sodium hypobromite.

Note

For the determination of small amounts of bromide by a catalytic method see Section 3.52.

3.29 Iodine

3.29.1 *Determination of iodine by reduction*

Both sodium thiosulphate and sulphur dioxide have been used as titrants for the thermometric determination of iodine. In twelve replicate titrations of $0\cdot1$ M I_2 in potassium iodide solution with M $Na_2S_2O_3$, using a rapid automatic titration apparatus, a relative standard deviation of $\pm 1\cdot3\%$ was obtained by Priestley and co-workers (1963a). The same workers (1968) used a continuous flow titrator for a similar titration. Priestley (1963a) also described the use of an automatic digital titrator and showed that with 2 M $Na_2S_2O_3$ as titrant extremely low concentrations of iodine of the order of $0\cdot001$ M, prepared by the addition of either sodium sulphite or bisulphite to potassium tri-iodide, could be titrated with relative standard deviations of less than $\pm 0\cdot5\%$. A much higher relative standard deviation of $\pm 1\cdot5\%$ was obtained in a similar titration of M/60 I_2 in potassium iodide. Thiosulphate solutions have a large endothermic heat of dilution but regardless of this disadvantage the precision of these methods is acceptable.

Hume and Duffield (1964) suggested that gaseous sulphur dioxide could be used as a titrant for the determination of iodine solutions.

3.29.2 *Determination of iodide by oxidation*

(*i*) OXIDATION WITH PERMANGANATE

Priestley and his co-workers (1963), using a rapid automatic titration apparatus, showed that 20 ml of dilute sulphuric acid containing 0·2 mmole of potassium iodide could be successfully titrated with 0·05 M $KMnO_4$ ($T = 1·3$ degC/mmole).

(*ii*) OXIDATION WITH CERIUM(IV)

Using an automatic digital titrator, Priestley (1963a) titrated M/120 KI with 0·25 M Ce(IV) in dilute sulphuric acid solution, and in eight titrations obtained a relative standard deviation of ±1·5%.

(*iii*) OXIDATION WITH HYPOCHLORITE

In the titration of 0·0125 M I^- by an incremental titration procedure (Brown *et al.*, 1969) seven titrations were shown to have an excellent precision, the relative standard deviation being only ±0·15%.

3.29.3 *Determination of iodide as silver iodide*

Harries (1966) reported the successful titration of iodide with silver nitrate, using a burette–thermopile apparatus, and Brown and co-workers (1969), using an incremental titration procedure, showed that a high precision could be obtained when 0·025 M I^- was titrated with 0·5 M $AgNO_3$, the relative standard deviation in this case being ±0·13%.

It is possible to titrate iodide in the presence of chloride but not in the presence of bromide, particularly at low concentration of either. The reasons for this are discussed under chlorine in Section 3.27.1.

3.29.4 *Determination of iodide and iodate by acid titration*

The reaction of hydrochloric acid with a mixture of iodide and iodate
$$2I^- + IO_3^- + 6H^+ = 3I^+ + 3H_2O$$
is exothermic and has been utilized by Priestley (1963a) to determine either iodide or iodate. Using an automatic digital titrator, either M/60 KI in excess of potassium iodate or M/360 KIO_3 in excess of potassium iodide was titrated with 1 M HCl and in five replicate titrations the relative standard deviation was ±0·8% for the iodide and, regardless of the sixfold dilution, ±0·2% for the iodate.

3.29.5 *Reaction of iodate with boron trifluoride*

Richards and Woolf (1969) determined the heat of reaction of boron trifluoride containing bromine with iodate and showed that potassium fluoride, iodine pentafluoride and oxygen were formed in the reaction.

Note

The titration of iodide with mercurous nitrate was not successful because of the disproportionation of the compounds formed initially

(Harries, 1968). The formation of complexes between iodine and organic sulphides is discussed in Section 4.13. For the determination of small amounts of iodide by a catalytic method see Section 3.52.

3.30 Copper

3.30.1 *Determination of copper(II) as the EDTA chelate*

Copper has a large heat of chelation with EDTA ($\Delta H° = -8\cdot3$ kcal/ mole) and as a consequence the reported methods of determination are sensitive and precise. Jordan and Alleman (1957) titrated $0\cdot0117$ M Cu^{2+} (25 ml) and obtained a precision of $0\cdot3\%$. They quoted a lowest level of determination of 1 mmole/l with an error of $\pm3\%$ ($T = 0\cdot33$ degC/ mmole). The thermometric method, therefore, has not the range of the volumetric method in which metallochromic indicators are used, because a similar error was obtained at the $0\cdot1$ mmole/l level by Wills *et al.*, (1962).

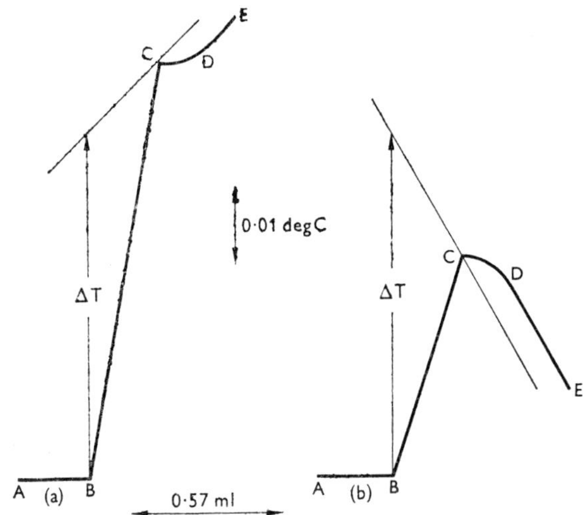

FIG. 3.4 Titration of copper(II) with M Na₄ EDTA.
(a), titrant warmer than titrand; (b), titrant cooler than titrand; from Jordan and Alleman (1957).

Jordan and Alleman's titration curve is shown in Fig. 3.4; the aberration at C–D is attributed to idiosyncratic behaviour of the thermistor.

Automatic apparatus has also been used for this determination and the results are summarized in Table 3.33.

The Na₄ EDTA used in all the titrations above was 1 M; the titration of copper in the presence of other ions is discussed in Section 3.49.

3.30.2 *Determination of copper(II) as* CuHg(CNS)₄

Chatterji (1958a) has determined copper in brass by titrating with $0\cdot4$ M $(NH_4)_2Hg(CNS)_4$, using the early titration method. The results were correct to within $0\cdot8\%$ when compared with the iodometric indicator method. Lead, tin and iron did not interfere with the method, but zinc, which did interfere, was allowed for by a second titration in which copper was first precipitated with ammonium thiocyanate. For further details see Section 3.31.1.

Table 3.33 Determination of copper(II) with EDTA

Cu^{2+} concentration	Replicates	Relative std. devn., %	Apparatus	Reference
0·10 M	11	±0·3	automatic	Priestley *et al.* (1963)
0·017 M	11	±0·5	automatic digital	Priestley (1963a)
Arbitrary	3	±1·0	continuous flow	Priestley *et al.* (1965, 1968)

3.30.3 *Determination of copper(II) iodometrically*

Thermometric titrimetry has been used in the well-known iodometric method for the determination of cupric ions (Billingham and Reed, 1964). Excess of iodide was added to the sample and after thermal equilibrium had been established the iodine formed was titrated with 1·2–2·5 M $Na_2S_2O_3$. A nitrogen atmosphere was maintained in the calorimeter to prevent oxidation. Copper concentrations between 50 and 2 mmole/l were determined with an average error of less than 3%. At lower concentrations the error became very large either because of the high endothermic heat of dilution of the titrant, estimated as 2·1 ± 0·5 kcal/mole, or because of the loss of iodine by sublimation The classical method with starch as indicator is much superior.

3.30.4 *Determination of copper(I) with dichromate*

Barthel and Schmahl (1965) have used a very precise thermometric titration apparatus to titrate 0·025 M CuCl in 200 ml of 1 M HCl with 0·5 N $K_2Cr_2O_7$. The results agreed within ±1·0% and because the enthalpy of the reaction was found to be high ($\Delta H = -33·5 \pm 1$ kcal/mole) a large temperature rise was obtained ($T = 0·55$ degC/mmole).

3.30.5 *Reaction of copper(II) with bases*

(*i*) DETERMINATION WITH CARBONATE

Priestley (1963a) obtained a relative standard deviation of ±0·8% in eight replicate titrations of M/60 $CuSO_4$ with 1 M Na_2CO_3, using an automatic digital titrator. The reaction in this case was endothermic.

(*ii*) REACTION WITH AMMONIA

The reaction of copper(II) salts with ammonia was first investigated by Dutoit and Grobet (1921) but no results were given. Siddhanta and Guha (1955) studied the formation of the cuprammines, using the early titration method, and also by direct injection. Ammonia solutions were titrated with copper sulphate solutions and the compound $[Cu(NH_3)_4]^{2+}$ was shown to be formed. The early titration method was used together with Job's method of continuous variation to determine the stoichiometry of the reactions (Siddhanta, 1948a, b).

(*iii*) REACTION WITH SODIUM HYDROXIDE

Haldar (1946b) investigated the basic sulphates of copper, using the early titration method. Titrations of solutions of copper(II) sulphate with sodium hydroxide solutions showed that the compounds $CuSO_4 \cdot CuO$, $NaCu(OH)_2$ and $Cu(OH)_3$ were formed.

Partially alcoholic solutions of the citrate and tartrate chelates of copper(II) have been titrated with 1 M NaOH (or NH_4OH if NaOH gave insoluble salts), using the early titration method (Ben-Yair and Jordan, 1951; Jordan and Ben-Yair, 1957). The titration of the copper(II) tartrate chelate showed it to be a diprotic acid whereas the citrate chelate was shown to be a monoprotic acid (T(tartrate) = 0·29 degC/mmole). The titration has been used as a method for the determination of citrates and tartrates (see Section 4.5).

3.30.6 *Reaction of copper with cyanides*

(*i*) FORMATION OF THE COPPER(I) COMPLEX CYANIDES

The enthalpy and entropy of the reactions
$$Cu^+ + 2CN^- \longrightarrow Cu(CN)_2^- + CN^- \longrightarrow Cu(CN)_3^{2-}$$
$$+ CN^- \rightarrow Cu(CN)_4^{3-}$$
have been determined by thermometric titration (Izatt *et al.*, 1967) and the stepwise instability constants of the complex ions $Cu(CN)_3^{2-}$ and $Cu(CN)_4^{3-}$ have been determined by Brenner (1965) as $1·0 \times 10^{-5}$ and $2·3 \times 10^{-3}$ respectively.

(*ii*) DETERMINATION OF COPPER(II) AS $[Cu(CN)_4]^{2-}$

The exothermic heat of formation of the copper(II) cyanide complex has been utilized by Sajó and Sipos (1968c) to determine the copper content of alloys containing tin, antimony and lead. The sample was dissolved in sulphuric and nitric acids, and 50% tartaric acid solution added.

Table 3.34 Determination of copper in alloys
(Comparison with known contents; from Sajó and Sipos (1968c))

Alloy	Known content %	Found %
GMS 67	67·30	67·1
G SO Ms-1	65·13	64·9
GZ Rg 5	84·45	84·1
LG Pb Sn 10	1·15	1·27
LG Sn 80	5·50	5·70

The solution was then neutralized with sodium hydroxide and a known excess added. The solution was diluted to 200 ml and injected with 8 ml of potassium cyanide titrant. The titrant was prepared by dissolving 500 g of potassium cyanide in 1 litre of water and adding 100 ml of 30% NaOH solution followed by 520 ml of glycerol. The titrant was expected to show no heat of dilution when 8 ml were injected into a blank solution, but if it did further cyanide or glycerol was added according to whether the temperature rose or fell. The results obtained by the method on samples of known copper content are given in Table 3.34.

(*iii*) HEXACYANOFERRATES(II) AND (III)

A number of workers have studied the formation of the hexacyanoferrate complexes of copper, using the early titration method. Aqueous or partially alcoholic solutions of the reactants were used: the concentration of the titrants was either 0·25 or 0·5 M and that of the titrand 0·1 M. With cupric sulphate as titrant and potassium hexacyanoferrate(III) as titrand and *vice versa*, the compound $Cu_3[Fe(CN)_6]_2$ was formed. (Gaur and Bhattacharya, 1950b), whereas with potassium hexacyanoferrate(II) the following compounds were formed: $K_2Cu[Fe(CN)_6]$ and $K_2Cu_3[Fe(CN)_6]_2$ (Bhattacharya and Gaur, 1947), and $K_2Cu_4[Fe(CN)_6]_3$ (Pâris and Le Chatelier, 1934).

3.30.7 *Formation of copper(II)–amine complexes*

A study of the copper–pyridine (Cu–Py) system has shown it to be a four-stage equilibrium producing the ions $[CuPy]^{2+}$, $[CuPy_2]^{2+}$, $[CuPy_3]^{2+}$ and $[CuPy_4]^{2+}$ (Izatt *et al.*, 1968). Approximately 1·5 M pyridine was used as titrant and 0·01 and 0·02 M $Cu(ClO_4)_2$ as titrand. A computer was used to evaluate the thermometric data from the titrations, giving the results shown in Table 3.35 (see also Christensen, Rytting and Izatt, 1969).

Table 3.35 Thermometric data for the copper–pyridine system at 25 °C
(From Izatt *et al.* (1968))

Reaction	$\log \beta_i$	$-\Delta H^\circ$ kcal/mole	$-\Delta S^\circ$ cal mole^{-1} degK^{-1}
$Cu^{2+} + \ \ Py \rightleftharpoons [CuPy]^{2+}$	$2{\cdot}50 \pm 0{\cdot}02$	$4{\cdot}02 \pm 0{\cdot}08$	$2{\cdot}0 \pm 0{\cdot}3$
$Cu^{2+} + 2Py \rightleftharpoons [CuPY_2]^{2+}$	$4{\cdot}30 \pm 0{\cdot}05$	$8{\cdot}86 \pm 0{\cdot}1$	$10{\cdot}0 \pm 0{\cdot}7$
$Cu^{2+} + 3Py \rightleftharpoons [CuPy_3]^{2+}$	$5{\cdot}16 \pm 0{\cdot}06$	$16{\cdot}11 \pm 0{\cdot}6$	$30{\cdot}6 \pm 2$
$Cu^{2+} + 4Py \rightleftharpoons [CuPy_4]^{2+}$	$6{\cdot}04 \pm 0{\cdot}1$	$21{\cdot}5 \ \pm 1{\cdot}5$	$41 \ \ \pm 5$

The enthalpy and entropy of successive steps in the formation of the copper(II) chelates of ethylenediamine and triethylenediamine have been determined by a direct-injection method (Poulsen and Bjerrum, 1955).

The elucidation of the structures of the copper(II) complexes with amino-acids is important because of their significance in enzyme reactions, and a number of these systems have been studied by thermometric titrimetry as shown in Table 3.36.

Table 3.36 Copper(II)–amino-acid complexes studied

Substance	Reference	Remarks
Glycine	Anderson *et al.* (1966a, b)	at 10, 25, 40 °C.
Diglycine Triglycine	}Brunetti *et al.* (1968)	at 25 °C, see Note
Tetraglycine	Nancollas and Poulton (1969)	see Note
Phenylalanine Alanine	}Anderson *et al.* (1966a, b)	at 10, 25, 40 °C

Note. The deprotonation of the ligand was also studied at high pH.

The thermodynamic properties of the chelation of copper(II) with tetracycline have been studied by Benet (1966), with 1, 2-diaminoethane, 1,3-diaminopropane, 4(5)-aminomethylimidazole, 4(5)-2-aminoethylimidazole, 2-aminomethylpyridine and 2,2'-aminoethylpyridine by Holmes and Williams (1967c) and with 1, 10-phenanthroline by Eatough (1970) who correlated the heat changes with the copper species produced. Details of Eatough's study are given in Section 3.31.6(*ii*).

3.30.8 *Ion-association in the copper(II) sulphate system*

In an investigation into the possible formation of anionic cupric sulphate complexes of the type $[Cu(SO_4)_2]^{2-}$ in concentrated solution, Becker and Grundmann (1969) added sodium perchlorate to the titrand (0·08 or 0·16 M $Cu(ClO_4)_2$) to obviate the large heat of dilution of the 2 M Li_2SO_4 titrant. No anionic complexes were found but the following results were obtained for the 1 : 1 ion-association: $K = 3·9 \pm 0·2$ l/mole; $\Delta H° = 1·74 \pm 0·10$ kcal/mole; and $\Delta S° = 8·5 \pm 0·4$ cal mole^{-1} degK^{-1}.

3.30.9 *Formation of copper(I) thiosulphate complexes*

The reaction between copper(II) and thiosulphate ions has been studied by Chatterji (1958c), using the early titration method. Breaks were observed in the titration curves at $Cu^{2+} : S_2O_3^{2-}$ ratios of 1 : 6, 1 : 3, 4 : 9 and 2 : 3 and were accompanied by the reduction of Cu(II) to Cu(I). The copper(I) complexes formed were interpreted as being $[Cu(S_2O_3)_5]^{9-}$, $[Cu(S_2O_3)_2]^{3-}$, $CuNa[Cu(S_2O_3)_2]$ and $[Cu_4(S_2O_3)_5]^{6-}$.

3.30.10 *Reaction of copper(II) with fluoride*

Sodium fluoride solutions have been titrated with copper sulphate solutions, both at various concentrations. Unlike similar titrations with Al^{3+} and Fe^{3+}, the titration curves did not show a sharp inflection on the formation of the $[CuF_4]^{2-}$ complex (Deschamps *et al.*, 1968).

3.30.11 *Formation of copper(II) hydroxy-acid chelates*

Ben-Yair and Jordan (1951) used the early titration method to determine the composition of copper(II) citrate and tartrate chelates and found that

1 : 1 complexes were formed, the citrate complex behaving as a monoprotic acid (see also Jordan and Ben-Yair, 1957).

Note

For the effect of copper on the determination of free acidity in uranyl sulphate see Section 3.48.1(*viii*) and for the determination of small amounts of copper(II) by a catalytic method see Section 3.52.

3.31 Zinc

3.31.1 *Determination of zinc as* ZnHg(CNS)$_4$

Chatterji (1949, 1958a) has determined zinc in brass by titrating with 0·4 M (NH$_4$)Hg(CNS)$_4$, using the early titration method; this reaction is one that has not been widely used in analytical chemistry. The brass (3 g) was dissolved in nitric acid and the nitric acid then removed with sulphuric acid. The resulting solution was neutralized with ammonia and brought to pH 4 with acetic acid in a final volume of 250 ml. Prior to the titration the copper present in a 50-ml aliquot was precipitated by the addition of 5 ml of saturated sulphur dioxide solution followed by 10 ml of 20% ammonium thiocyanate solution. The titration was stated to be slow and the results to be within 1% of those obtained gravimetrically ($T = 0·20$ degC/mmole).

3.31.2 *Determination of zinc as the* EDTA *chelate*

Jordan and Alleman (1957) titrated 0·01 M Zn^{2+} with 1 M Na$_4$EDTA and obtained a precision of 0·7% and an error of 0·8%. They quoted a lowest level of determination of 2 mmole/l with an error of 3%. A relative standard deviation of $\pm0·4\%$ was obtained in five replicate titrations of M/60 ZnSO$_4$ with M Na$_4$ EDTA, in an automatic digital titrator (Priestley, 1963a). These precisions are not as good as those obtained in a volumetric procedure (cf. Brown and Hayes, 1953).

3.31.3 *Reaction of zinc with complex iron cyanides*

(*i*) COMPOSITION OF ZINC HEXACYANOFERRATES(II) AND (III)

The composition of these compounds has been studied by the early titration method. Aqueous or partly alcoholic solutions of the reactants were used; the concentrations of the titrants were 1·0 M and those of the titrands 0·1 M. With zinc sulphate as titrant and potassium hexacyanoferrate(III) as titrand and *vice versa*, the compound Zn$_3$[Fe(CN)$_6$]$_2$ was formed (Gaur and Bhattacharya, 1952a); with potassium hexacyanoferrate(II) the compound ZnK$_2$[Fe(CN)$_6$] was formed (Pâris and Le Chatelier, 1934).

(*ii*) DETERMINATION OF ZINC AS ZnK$_2$[Fe(CN)$_6$]

A simple automatic titration apparatus has been used to determine zinc by titration with potassium hexacyanoferrate(II). Both direct and differential titrations were employed, the latter giving a sharper end-point. (Takeuchi and Yamazaki, 1969a).

3.31.4 *Reaction of zinc with cyanides*

(*i*) FORMATION OF ZINC CYANIDE COMPLEXES

The early titration method has been used to study the formation of zinc cyanide complexes. With potassium cyanide as titrant the compounds $Zn(CN)_2$ and $K_2[Zn(CN)_4]$ were shown to be formed (Mondain-Monval and Pâris, 1934).

In recent work by Izatt and his co-workers (1971) the $\Delta H°$ values at 20°, 25° and 40 °C have been reported for the formation of the zinc cyanide complexes $Zn(CN)_n$, where $n = 2, 3$ or 4. A dilute titrand (4 mM) was used to prevent the precipitation of $Zn(CN)_2$.

(*ii*) DETERMINATION OF ZINC AS $K_2[Zn(CN)_4]$

Sipos and Sajó (1966) have determined the zinc content of manganese–zinc ferrites, using a direct-injection method. The titrant, potassium cyanide, has to be added in large excess to ensure that the reaction is complete, as the zinc cyanide complex has a low stability constant.

The ferrite was taken into solution with hydrochloric acid, and to mask iron and manganese, tartaric acid and EDTA were added. The solution was neutralized with sodium hydroxide until it became a clear green and was then titrated with (presumably) 2 ml of 50% potassium cyanide solution. The results obtained by the method are given in Table 3.37.

Table 3.37 Determination of zinc in ferrites
(Comparison of methods; from Sipos and Sajó (1966))

Sample	Complexometric method, %	Enthalpimetric method, %
1	12·1	12·3
2	11·9	11·8
3	13·4	13·3
4	6·3	6·4
5	9·7	9·7
6	14·1	13·9

3.31.5 *Reaction of zinc salts with bases*

(*i*) FORMATION OF BASIC ZINC SALTS

Dutoit and Grobet (1921) studied the basic salts of zinc nitrate by the early titration method. They titrated 25 ml of 0·4 N $Zn(NO_3)_2$ with 6 M NaOH and showed that the compounds $ZnNO_3·OH$, $Zn(OH)_2$ and $Na_2[Zn(OH)_4]$ were formed successively: Ben-Yair (1961) reported the formation of only the first compound. A similar study by Haldar (1946c) with zinc sulphate as titrant showed that the compounds $ZnSO_4·3Zn(OH)_2$, $ZnSO_4·Zn(OH)_2$, $Zn(OH)_2$ and $Na_2[Zn(OH)_2]$ were formed.

(ii) DETERMINATION OF ZINC WITH BASES

Ben-Yair (1957) reported the determination of zinc by thermometric titration with ammonia or sodium hydroxide, using the early titration method but later (1961) stated that the estimation of zinc by titration with ammonia was not feasible. A relative standard deviation of $\pm 1.1\%$ was obtained in seven replicate titrations of $M/60$ $Zn(NO_3)_2$ with 1 M Na_2CO_3, in an automatic digital titrator (Priestley, 1963a). The reaction in this case was endothermic.

(iii) CITRATE AND TARTRATE CHELATES OF ZINC

Partially alcoholic solutions of the citrate and tartrate chelates of zinc have been titrated with 1 M NaOH (or NH_4OH if the NaOH gave insoluble salts) using the early titration method (Jordan and Ben-Yair, 1957). The titration of the zinc tartrate chelate (200 ml) gave a curve with a single inflection which showed it to be a diprotic acid ($T \sim 0.001$ degC/ mmole) whereas the zinc citrate chelate was shown to be a monoprotic acid. This titration has been used as the basis of a method for the determination of citrates and tartrates (see Section 4.5). In an earlier paper the same authors described the thermometric determination of the composition of these compounds and found them to be 1 : 1 chelates (Ben-Yair and Jordan, 1951).

3.31.6 *Formation of zinc(II) complexes*

(i) ZINC(II) HALIDE COMPLEXES

Gerding (1969) determined the enthalpy changes for the stepwise formation of zinc(II) halide complexes. Successive small quantities of the 1 M sodium salts of each halide were added to 1 M $Zn(ClO_4)_2$ in 1 M $NaClO_4$ solution or *vice versa* to reduce heat of mixing and it was found that ΔH becomes less positive in value from fluoride to bromide and probably also to iodide.

Table 3.38 Thermodynamic data for the reaction of zinc(II) and copper(II) with 1,10-phenanthroline

(From Eatough, 1970)

Reaction	Log K	$-\Delta H°$ kcal/mole	ΔS cal mole^{-1} degK^{-1}
$Zn^{2+} + P = ZnP^{2+}$	6.17 ± 0.1	7.5 ± 0.2	3.1 ± 0.6
$Cu^{2+} + P = CuP^{2+}$	9.14 ± 0.06	11.03 ± 0.1	4.8 ± 0.3
$ZnP^{2+} + P = ZnP_2^{2+}$	5.91 ± 0.1	4.76 ± 0.1	11.1 ± 0.4
$CuP^{2+} + P = CuP_2^{2+}$	6.87 ± 0.08	5.42 ± 0.1	13.2 ± 0.3
$ZnP_2^{2+} + P = ZnP_3^{2+}$	5.25 ± 0.1	2.9 ± 0.7	14.3 ± 2
$CuP_2^{2+} + P = CuP_3^{2+}$	5.42 ± 0.1	5.1 ± 0.3	1.1 ± 0.9

(ii) ZINC(II) 1,10-PHENANTHROLINE COMPLEX

Eatough (1970) studied the formation of the very stable complexes of 1,10-phenanthroline (P) with zinc(II) and copper(II), having the form MP^{2+}, MP_2^{2+} and MP_3^{2+}. Two different metal nitrate solutions containing sufficient perchloric acid to prevent hydrolysis were titrated with

P–nitric acid solution, and the temperature rises were corrected for the effects of the nitric and perchloric acids. The results obtained are shown in Table 3.38.

Note

For the determination of diethylzinc see Section 3.50, for the behaviour of zinc salts in an investigation of ligand number see Section 3.51. and for the determination of small amounts of zinc by a catalytic method see Section 3.52.

3.32 Titanium

The reaction properties and determination of titanium have largely been concerned with the oxidation of titanium from its lower valency states. Jordan and Ewing (1960) determined the enthalpy of the cerium(IV) oxidation of titanium(III) and obtained the high value of -30 ± 1 kcal/eq, but no analytical results were given in the paper. Ewing (1961) stated that iron(II) and titanium(III) in admixture could be titrated in a single titration with cerium(IV), two distinct inflections being shown in the titration curve. He stated that titrations of this type are possible whenever the successive free energies of the reaction differ by more than 3–5 kcal/eq.

3.32.1 *Determination of titanium(III) by oxidation with chromium(VI)*

Barthel and Schmahl (1965) used a precise thermometric titration apparatus in the titration of 0·01 M $TiCl_3$ with M $K_2Cr_2O_7$ in 1 M HCl solution. The results were precise to $\pm 0·1\%$ with no bias, which is an indication of the high heat of reaction, determined in the titration as -36 ± 2 kcal/eq. At the same time the entropy of the reaction was found to be $+7 \pm 3$ cal^{-1} eq^{-1} degK^{-1}. Well-defined end-points and an accuracy that compares favourably with the visual back-titration method with permanganate after the addition of excess of ferrous ions make this method very acceptable.

3.32.2 *Reaction of titanium(IV) with hydrogen peroxide*

(*i*) FORMATION OF TITANIUM PEROXY SALTS

The reaction of titanium(IV) salts with hydrogen peroxide, producing a yellow colour, which is the basis of the colorimetric method for titanium, has been studied by Rivenq (1954b), using the early titration method. A solution of 23·3 mmole of $TiCl_4$ in 100 ml of hydrochloric acid was titrated with 1·76 N H_2O_2 and the titration curve had a sharp inflection at the end-point. The titration showed that the exothermic reaction could be represented by $TiO_2 + H_2O_2 \rightarrow TiO(O_2) + H_2O$ ($T = 0·09$ degC/mmole).

(*ii*) DETERMINATION OF TITANIUM(IV) WITH HYDROGEN PEROXIDE

Hungarian workers have used the exothermic titanium(IV)–hydrogen peroxide reaction to determine titanium in a wide range of materials by a direct-injection method. The samples were brought into solution as described under Iron (see Section 3.36.2(*i*)) and boric acid was added to complex any fluoride present and thus leave the titanium ions free to react with the hydrogen peroxide. Mercuric chloride was added to remove interfering sulphides and the solution was then oxidized with ammonium persulphate. Alternatively, the solution after the determination of iron was used. Finally the solution was injected with 2 ml of 30% hydrogen peroxide. A linear calibration graph was obtained for titanium contents up to a concentration of 0·5 g/l and at this level a ΔT value of 0·07 degC was observed.

The method has been used to determine the titanium content of slags, clinkers and cements, with the results given in Table 3.39.

Table 3.39 Determination of % titanium in slags, clinkers and cements
(Comparison of results; from Sajó and Sipos (1966, 1968a))

Sample		Colorimetric method	Enthalpimetric method
Blast-furnace slags	1	0·32	0·3
	2	0·37	0·4
	3	0·27	0·3
	4	0·66	0·7
Siemens–Martin slags	1	0·44	0·4
	2	0·36	0·4
Clinkers	1	0·57	0·6
	2	0·30	0·3
	3	0·25	0·3
Cements	1	0·22	0·2
	2	0·34	0·3

No results were given for the analysis of silicates (Sajó, 1969c) but those for clays and ores, in which the analysis was made on 1- or 2-g samples, are given in Table 3.40 (Sajó and Sipos, 1968b; 1969b). It can be seen that the results with the larger samples are somewhat more precise.

Table 3.40 Determination of titanium as % TiO_2 in clays and minerals
(Comparison with known contents, using two sample weights; from Sajó and Sipos (1968b))

Sample	Known content	Found	
		1 g taken	2 g taken
Fireclay 1	1·85	2·10	1·80
Fireclay 2	1·90	2·10	1·80
Fireclay 3	1·40	1·60	1·40
China clay	0·09	0·2	0·10
Felspar	0·02	0·1	0·1
Granite	0·3	0·5	0·3

Vanadium interferes with the determination but can be allowed for by using a correction factor based on the vanadium content. This can be determined by direct injection (see Section 3.33.1(ii)).

Taubinger (1969) has used the dual titration method to determine titanium(IV), by titration with titanium(IV) and using hydrogen peroxide as the reactant.

3.33 Vanadium

3.33.1 *Reaction of vanadium*(V) *with hydrogen peroxide*

(i) FORMATION OF VANADIUM PEROXY SALTS

The reaction of vanadium(V) with hydrogen peroxide, producing a reddish-brown colour, which is the basis of a colorimetric method for vanadium, has been studied by the early titration method (Rivenq, 1945a). A solution of 3·70 mmole of V_2O_5 in 200 ml of 5 N H_2SO_4 was titrated with 0·374 M hydrogen peroxide. The titration curve had a sharp end-point which showed that the exothermic reaction could be represented by $V_2O_5 + 2H_2O_2 \rightarrow V_2(O)_3(O_2)_2 + 2H_2O$ ($T = 0·12$ degC/mmole). It was found that if the acid concentration were not high enough reduction of the vanadium took place instead of the peroxide substitution reaction. These results were confirmed by potentiometric titration but did not agree with those of Thiesse (1940) who postulated the formation of $V_2O_5 \cdot 2H_2O_2$.

(ii) DETERMINATION OF VANADIUM(V) WITH HYDROGEN
  PEROXIDE

The reaction of vanadium(V) with hydrogen peroxide described above has been utilized by Sajó (1969c) to determine the vanadium content of silicates by the direct-injection method. Vanadium can be determined in the presence of titanium with hydrogen peroxide providing fluoride is present: vanadium does not complex with the fluoride ions as does titanium, and the titanium fluoro-complexes do not react with hydrogen peroxide. The solution should also be acid to prevent the reduction reaction (see (i) above).

The sample was prepared in the same manner as described for titanium (see Section 3.32.2(ii)) except that hydrofluoric acid was added if it had not been used in the dissolution of the sample, and no boric acid was needed. Unfortunately no results are available.

Taubinger (1969b) has used the dual titration method to determine vanadium(V) in the presence of titanium(IV) ions, by titration with vanadium(V) solution with hydrogen peroxide as the reactant (see Fig. 1.9).

3.34 Chromium

3.34.1 *Reactions of chromium*(VI)

(i) DETERMINATION OF CHROMIUM(VI) BY REDUCTION WITH
  IRON(II)

The iron(II) reduction of dichromate has a high heat of reaction (see Section 3.36.1(ii)) and the thermometric titrations have a better precision and accuracy than the similar volumetric titration with diphenylamine as indicator (Barthel and Schmahl, 1965). Sajó and Sipos (1967a) have determined the chromic acid content of chromium-plating solutions. A portion of the plating solution was diluted to give a solution containing about 40 g/1 of H_2CrO_4, then 5 ml of this diluted solution were placed in the reaction beaker, 50 ml of concentrated sulphuric acid were added, the contents diluted to the 200-ml mark and then injected with 5 ml of iron(II) solution which had zero heat of dilution. The titrant was prepared by dissolving 200 g of $FeSO_4 \cdot 7H_2O$ in 440 ml of water and adding 150 ml of concentrated sulphuric acid. The titrant was injected into a blank solution, and if the temperature rose or fell either iron(II) sulphate or sulphuric acid respectively was added until no heat of dilution occurred.

The results obtained for various chromium-plating solutions are given in Table 3.41 from which it can be seen that the results differ from those of the indicator method by less than $\pm 0.7\%$.

Table 3.41 Chromium content as g/1 H_2CrO_4 in plating solutions
(Comparison of methods; from Sajó and Sipos (1967a))

Sample	Indicator method	Enthalpimetric method
1	435·2	434·0
2	397·4	399·0
3	321·3	321·0
4	169·2	168·0
5	157·3	158·0
6	158·1	158·0

The same method has been used to determine the chromium content of slags but ascorbic acid can also be used as the reductant, as described in the next section.

(*ii*) DETERMINATION OF CHROMIUM(VI) BY REDUCTION WITH ASCORBIC ACID

The chromium content of slags has been determined by oxidative dissolution followed by direct injection with iron(II) as described above or with ascorbic acid (Sajó and Sipos, 1966). The dissolution of the slag with the mixture described in the determination of aluminium (see Section 3.10.2) resulted in the oxidation of the chromium to Cr(VI). The solution (200 ml) was injected with either iron(II), as described above, or with ascorbic acid prepared by dissolving 15 g of the acid in 100 ml of water and adding 5 ml of concentrated sulphuric acid, 20 ml of perchloric acid, 60 ml of hydrochloric acid and 20 ml of phosphoric acid, i.e., adding the acid mixture used for the dissolution of the slag. This solution, though an effective reductant for chromium, was stated to have a shorter shelf-life than the iron(II) solution. The results obtained on a number of blast-furnace slags are given in Table 3.42.

Table 3.42 Determination of chromium as % Cr_2O_3 in slags
(Comparison of methods; from Sajó and Sipos (1966))

Sample		Potentiometric method	Enthalpimetric method
Blast-furnace slag	1	1·86	1·7
	2	1·12	1·2
	3	0·89	0·9
	4	0·12	0·2

(*iii*) REACTION OF CHROMIUM(VI) WITH SILVER NITRATE

Harries (1966) has titrated potassium dichromate with silver nitrate, using a burette–thermopile apparatus, and the formation of silver chromate was indicated and verified by analysis.

3.34.2 *Determination of chromium(III)*

(*i*) DETERMINATION OF CHROMIUM(III) AS THE EDTA CHELATE

Priestley (1963a), using an automatic digital titrator, has titrated $M/60$ $Cr(NO_3)_2$ with 1 M Na_4EDTA and in seven replicate titrations obtained a relative standard deviation of $\pm0.2\%$. The reaction in this case was endothermic.

(*ii*) DETERMINATION OF CHROMIUM(III) BY OXIDATION

Sajó and Sipos (1967a) have determined the chromium(III) content of chromium-plating solutions by permanganate oxidation. A 10-ml portion of the plating solution was placed in the titration vessel, 30 ml of 30% NaOH solution were added, the solution was diluted to the 200-ml mark

Table 3.43 Chromium(III) content in plating solutions
(Comparison of methods; from Sajó and Sipos (1967a))

Sample	Volumetric method, g/l	Enthalpimetric method, g/l
1	3·92	3·98
2	3·63	3·71
3	2·94	2·90
4	1·72	1·67
5	1·35	1·40
6	0·67	0·68

and 5 ml of potassium permanganate titrant were then injected. The permanganate titrant was expected to show no heat of dilution when injected into a blank solution (i.e., 30 ml of 30% NaOH solution diluted to 200 ml). Either sodium chloride or potassium permanganate solution was added to the solution until no heat of dilution occurred, the former if the temperature rose, the latter if it fell. The results given in Table 3.43 show that even at these low levels they are well within $\pm 4\%$ of those from a volumetric method.

3.34.3 *Formation of chromium peroxy salts*

When hydrogen peroxide is added to a chromate solution a blue colour is produced owing to the formation of peroxy salts. This reaction has been studied thermometrically by Rivenq (1945a) using the early titration method. A solution of 50 ml of $0\cdot303$ N H_2O_2 was titrated with $0\cdot474$ N K_2CrO_4 and a sharp end-point was obtained which showed that this highly exothermic reaction could be represented by $CrO_4 + 2H_2O_2 \rightarrow CrO(O_2)_2 + 2H_2O$ ($T = 0\cdot74$ degC/mmole).

Note

For the thermodynamic properties of silver and lead chromates see Sections 3.39.8 and 3.17.9.

3.35 Manganese

3.35.1 *Determination of manganese by oxidation of* Mn(II) *to* Mn(IV)

Hungarian workers have determined manganese in a wide variety of materials by a direct-injection method, the principle of which is the exothermic oxidation of manganese(II) ions by potassium permanganate in

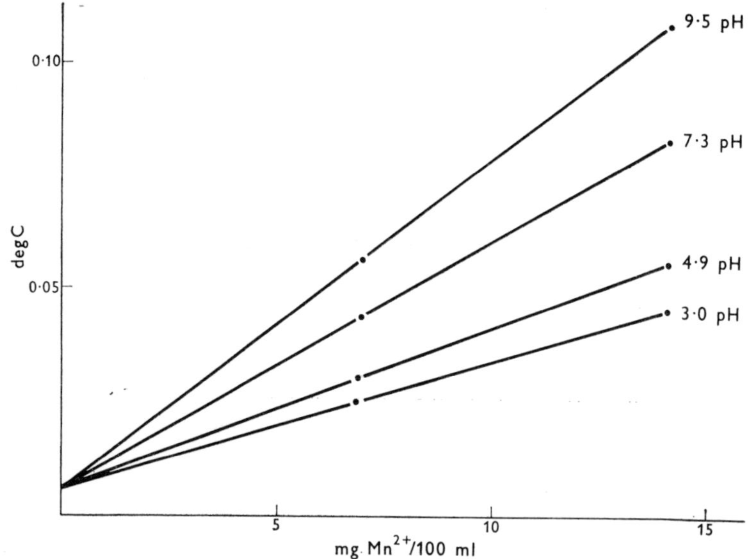

Fig. 3.5 Relation between temperature rise and pH in enthalpimetric titration of 14 mg of Mn(II) in 100 ml of solution with saturated KMnO$_4$ solution (Sajó, 1966b).

alkaline solution containing an excess of ammonium ions (Sajó, 1966b).

The sample was brought into solution with acids or alkali fusion so that all the manganese present was in the manganese(II) state and the iron in the iron(II) state; or alternatively the solution remaining after the titration of silica (see Section 3.15.1) was used. The sample solution was diluted to 100 ml with 10% ammonium chloride solution, then neutralized with 35% ammonia solution until iron etc. had precipitated and a further 5 ml were added to bring the pH to between 7·5 and 9 (see Fig. 3.5). The solution was cooled, diluted to 200 ml and titrated immediately with 5 ml of saturated potassium permanganate solution (this solution has no heat of dilution).

The method has been applied by Sajó and Sipos to slags (1966), cements and clinkers (1968a, 1969b), magnesites and dolomites (1968d), and by Sajó to silicates (1969c) provided that no sample contained less than 1% of manganese. If the content were lower than 1% the reduction method was used (see below). The results obtained by the method are compared with the Volhard–Wolff volumetric method in Table 3.44.

Table 3.44 Determination of manganese as % MnO in slags and cements (Comparison of methods; from Sajó and Sipos (1966, 1968a))

Sample		Volumetric method	Enthalpimetric method
Blast-furnace slag	1	2·05	2·1
	2	1·83	1·9
	3	1·83	2·0
Siemens–Martin slag	1	5·08	5·2
	2	10·2	9·8
	3	4·54	4·6
	4	11·1	11·4
Cement	1	1·8	1·9
	2	2·3	2·1

The solution remaining after the titration of manganese can be used for the determination of calcium and magnesium (see Section 3.6.2(*ii*)).

3.35.2 *Determination of manganese by reduction of* Mn(VII) *to* Mn(II)

Sajó and Sipos have also determined manganese by reduction of permanganate to manganese(II). This reduction reaction is more exothermic than the oxidation reaction above. They suggested that the reduction method should be used when the manganese content was in the range 0·01–1·0% and applied it to the analysis of silicates, using formaldehyde as the reductant (Sajó and Sipos, 1969c), and to the analysis of zinc–manganese ferrites, using hydrogen peroxide as the reductant (Sajó and Sipos, 1966).

(i) DETERMINATION WITH FORMALDEHYDE

The sample was brought into solution as described in the oxidation method above, oxidized with ammonium persulphate (silver nitrate as catalyst) and boiled. After cooling, the solution was titrated with 3% formaldehyde solution. Unfortunately no results are available for low manganese contents, for which this method is recommended.

(ii) DETERMINATION WITH HYDROGEN PEROXIDE

The sample was brought into solution as described in the oxidation method (Section 3.35.1) and to remove chloride either silver nitrate was added or the solution was evaporated after addition of sulphuric acid. The solution was oxidized with ammonium persulphate, and 0·5% silver nitrate solution was added as a catalyst. It was then boiled and after cooling 2 ml of 30% hydrogen peroxide were injected (Sipos and Sajó, 1966). The results obtained by the method are given in Table 3.45.

Table 3.45 Determination of manganese in ferrites
(Comparison of methods; from Sipos and Sajó (1966))

Sample	Complexometric method, %	Enthalpimetric method, %
1	15·3	15·1
2	14·9	15·0
3	12·9	12·9
4	18·2	18·4
5	16·6	16·5
6	15·7	15·6

These results seem to have a better accuracy than those obtained by the oxidation method above.

3.35.3 *Determination of manganese(II) as the* EDTA *chelate*

A continuous-flow titrator was used to titrate manganese(II) sulphate with M Na_4EDTA and results correct to within 1% were obtained (Priestley *et al.*, 1965). Attempts to titrate manganese(II) chloride with M Na_4EDTA, using an automatic digital titrator, were unsuccessful (Priestley, 1963a).

3.35.4 *Determination of manganese(VII)*

Harries (1966) used a burette–thermopile apparatus in the determination of permanganate. With 0·5 M $K_4[Fe(CN)_6]$ as titrant he titrated acidified 0·002 M $KMnO_4$ and obtained a precision of $\pm0·6\%$. With mercury(II) nitrate as titrant no results were obtained, as the reaction was not stoichiometric.

Note

For the determination of small amounts of manganese by a catalytic method see Section 3.52.

3.36 Iron

3.36.1 *Determination of iron(II) by oxidation*

(*i*) OXIDATION OF IRON(II) WITH CERIUM(IV) SULPHATE

The ceric oxidation of iron(II) ions was shown (Jordan and Ewing, 1960; Ewing, 1961) to have a high heat of reaction ($\Delta H^\circ = -24\cdot2$ kcal/mole; see Table 3.67) and because of this a precision and error of $\pm 1\%$ and $\pm 2\%$ could be obtained at concentration levels of 1 mM and 0·1 mM respectively. In the assessment of a differential titration procedure Tyson and co-workers (1961) titrated between 0·1775 and 0·3853 meq of $FeSO_4$ in 50–75 ml of solution with 0·3483 M $Ce(SO_4)_2$ in 3 M H_2SO_4. The error in twelve titrations was within the range from $+0\cdot9$ to $-1\cdot4\%$ with an average of $-0\cdot4\%$.

Ewing (1960) and Jordan (1961) both showed that iron(II) and titanium(III) in a binary mixture could be determined separately in a single titration because the titration curve showed a distinct inflection marking the end-point between the two oxidations. They stated that titrations of this type are possible whenever the successive free energies of the reactions differ by more than 3–5 kcal/eq.

(*ii*) OXIDATION OF IRON(II) WITH DICHROMATE

This oxidation has an even higher heat of reaction than the cerium oxidation above. Jordan and Ewing (1960) obtained a value for ΔH° of -27 ± 1 kcal/eq in an unspecified strength of sulphuric acid. Barthel and Schmahl (1965) obtained a value of $-24\cdot1 \pm 0\cdot5$ kcal/eq in 0·5 M H_2SO_4, using a precise thermometric titration procedure in which they titrated 0·002 M $(NH_4)_2Fe(SO_4)_2$ with 0·04 N $K_2Cr_2O_7$ and obtained accurate results with a precision of $\pm 0\cdot3\%$, whereas a similar volumetric titration using diphenylamine as indicator gave results precise to within $\pm 0\cdot5\%$ and with a bias of $+0\cdot2\%$. Similar results were obtained in the reverse titration. Takeuchi and Yamazaki (1969a) obtained an error and precision of less than 1% and 3% respectively when they titrated 0·003 M Fe^{2+} with dichromate.

(*iii*) OXIDATION OF IRON(II) WITH PERMANGANATE

This oxidation reaction has a similar heat of reaction to the dichromate oxidation above: $\Delta H^\circ = -27 \pm 1$ and $-25\cdot3 \pm 0\cdot5$ kcal/eq, obtained by Jordan and Ewing (1960) and Barthel and Schmahl (1965) respectively. Mayr and Fisch (1929), using the early titration method, stated that they needed to use only 0·05 N $KMnO_4$ for the oxidation of iron(II) in sulphuric acid medium which was fortunate in view of the low solubility of permanganate. The same concentration of permanganate was used in an automatic titration procedure in which 20 ml containing 0·2 mmole of $FeSO_4$ were titrated in dilute sulphuric acid. A large T value of 1·4 degC/mmole was obtained (Priestley *et al.*, 1963).

In obtaining the ΔH° value above, Barthel and Schmahl titrated 0·002 M $(NH_4)_2Fe(SO_4)_2$ with 0·04 N $KMnO_4$ in 0·5–1·0 N H_2SO_4 and the results were precise to within $\pm 0.3\%$ with no error, whereas the self-indicating volumetric titration results agreed to within $\pm 0.5\%$ and had a bias of -0.2%. Although his results were not very good Müller (1941) used the iron(II)–permanganate reaction to demonstrate his use of the multi-junction thermopile temperature-detector.

(*iv*) OXIDATION OF IRON(II) WITH HYDROGEN PEROXIDE

Sajó (1957), in a paper describing the determination of the basicity of slag by dissolution of it in hydrochloric acid, also described the determination of the ferrous iron content by the direct injection of hydrogen peroxide immediately after the dissolution heat had stabilized. This procedure was criticized by Mandl and co-workers (1962) who found that a secondary reaction, $HCl + H_2O_2 \rightarrow Cl_2$, catalysed by ferric ions, was interfering. They reversed the order, first determining the silica content by injection of hydrogen fluoride (see Section 3.15.1) and then injecting the hydrogen peroxide. In this case iron(III) ions were present as their fluoride complex which has no catalytic effect on the $HCl–H_2O_2$ reaction (see also Sajó, 1969d). This method obviously suffers from interference from titanium (see Section 3.32.2(*ii*)).

(*v*) OXIDATION OF IRON(II) WITH PERSULPHATE

Sajó and Sipos have used the persulphate oxidation of iron(II) to iron(III) to determine both the iron(II) and the iron(III) content, the latter being determined after being reduced to iron(II). The details of the method are given in the next section and the results of the determination of iron(II) in slags are given in Table 3.46 and agree well with those obtained volumetrically.

Table 3.46 Determination of iron(II) as % FeO in slags
(Comparison of methods; from Sajó and Sipos (1966))

Sample		Volumetric method	Enthalpimetric method
Siemens–Martin slag	1	11·05	11·3
	2	15·46	15·6
	3	10·18	10·5
	4	13·27	13·1
Blast-furnace slag	1	0·37	0·3
	2	0·46	0·5
	3	0·42	0·5
	4	1·23	1·2

3.36.2 *Determination of iron(III)*

(*i*) DETERMINATION OF IRON(III) BY REDUCTION–OXIDATION

The persulphate oxidation method outlined above has been used by Sajó and Sipos to determine the total iron content of a wide variety of substances. The samples were brought into solution with hydrochloric or hydrofluoric acids, or by alkali fusion, and the solution was acidified with hydrochloric acid if necessary. Alternatively, the solution remaining after the determination of aluminium, barium or silica was used. If the solution contained fluorides, e.g., after the determination of aluminium (see Sections 3.10.2 and 3.10.3), boric acid was added to complex all the fluoride as fluoroborate and release the complexed Fe^{3+} ions for the reduction stage.

The main reductant used was 0·5 ml of a 30% solution of sodium poly-sulphide except in the analysis of clays (Sajó and Sipos, 1968b) when stannous chloride was used (35 g in 400 ml of conc. HCl diluted to 1 litre). In all cases the excess of reductant and sulphides was removed by the addition of 10 ml of saturated mercuric chloride solution. The solution, after dilution to 200 ml, was injected with 8 ml of a 200 g/l solution of ammonium persulphate. The titrant was expected to have no heat of dilution when 8 ml were injected into a blank solution, but if this were not so the titrant was diluted or made more concentrated according to whether the temperature rose or fell, until no heat of dilution occurred.

The method was used by Sajó and Sipos to determine the total iron contents of slags (1966), cements and clinkers (1968a), clays (1968b, 1969b), magnesites and dolomites (1968d), and of silicates (Sajó, 1969c). A series of twelve determinations of the total iron content of a standard magnesite known to contain 4·22% Fe_2O_3 had a range of 4·13–4·34% with an average of 4·25%.

Table 3.47 Determination of iron as % Fe_2O_3 in slags, cements, clinkers, magnesites and dolomites
(Comparison with known contents; from Sajó and Sipos (1968a))

Sample		*Known*	*Found*
Siemens–Martin slags	1	13·96	13·7
	2	14·10	14·1
	3	9·35	9·5
	4	14·24	14·4
Blast-furnace slags	1	0·32	0·4
	2	0·40	0·5
	3	0·39	0·4
	4	0·92	0·9
Clinkers	1	1·8	1·9
	2	2·17	2·4
	3	2·75	2·8
Cements	1	2·34	2·2
	2	2·02	2·0
Magnesites	O	4·22	4·13
	K	3·72	3·61
	VK	3·70	3·80
Dolomite		1·62	1·80
Burnt dolomite		2·10	2·18

The results obtained in the analysis of slags, cements, clinkers, magnesites and dolomites for their total iron content are given in Table 3.47.

In the analysis of various types of clays and minerals, Sajó and Sipos (1968b) determined the iron content on samples of 1 and 2 g, with the results given in Table 3.48 from which it can be seen that the accuracy of the method is improved when the higher sample weight is used.

The solution remaining after the determination of iron can be used for the determination of titanium (see Section 3.32.2(*ii*)).

Table 3.48 Determination of iron as % Fe_2O_3 in clays and minerals
(Comparison with known contents using two sample weights; from Sajó and Sipos (1968b))

| *Sample* | *Known content* | *Found* | |
		1 g *taken*	2 g *taken*
Fireclay 1	2·06	2·20	2·05
Fireclay 2	1·74	1·90	1·70
Fireclay 3	1·12	1·30	1·10
China clay	1·30	1·10	1·35
Felspar	0·06	0·10	0·10
Granite	1·90	1·90	1·85

A similar method to that described above has been used to determine the iron content of nickel ores and of sintered nickel oxides. The sample (2·0 g) was taken up in hydrofluoric and nitric acids, hydrochloric acid was added and the solution evaporated, filtered to remove silica and diluted to 200 ml. To reduce iron(III) to iron(II), 25 ml of acetate buffer and 50 ml of saturated hydrogen sulphide solution were added to an aliquot (up to

Table 3.49 Determination of iron in nickel ores
(Comparison of methods; from Rády *et al.* (1967a))

Ore source	*Gravimetric method*, %	*Enthalpimetric method*, %
Cuba	0·48	0·42
Canada	0·40	0·42
Soviet Union	0·58	0·60

100 ml) containing 5–20 mg Fe. Finally the solution was diluted to 200 ml, 25 ml of saturated mercuric chloride solution were added to remove the excess of reductant and sulphides; after thermal equilibrium had been established, 5 ml of 10% $(NH_4)_2S_2O_8$ solution were injected.

In this case the acetate buffer of pH 3·6 gave a calibration graph for iron *vs.* temperature-rise that was linear and passed through the origin, i.e., there was no heat of dilution of the titrant. Potassium chlorostannite was tried as a reductant, but though it was effective it interfered with the subsequent determination of nickel and of cobalt. It was found that 10 mg of iron could be determined to within $\pm 4\%$ even in the presence of 1000 mg of nickel and 800 mg of cobalt. Table 3.49 gives the results obtained by the method on a series of nickel ores.

(*ii*) DETERMINATION OF IRON(III) AS THE EDTA CHELATE

Very little work has been done on the thermometric titration of iron with EDTA regardless of the fact that at pH 1·8 the EDTA titration of iron with salicylic acid as indicator is almost specific (Radmacher and Schmitz, 1959.) An automatic thermometric titration has been used to titrate a mixture of iron, aluminium and other metals with M Na_4EDTA. The titration curve ($T = 0.08$ degC/mmole) showed a distinct inflection because the reaction of aluminium, titrated after iron, is endothermic (Priestley, *et al.*, 1963). With an automatic digital titrator no results were obtained in a titration of M/60 $Fe(NO_3)_2$ or of M/60 $Fe(NO_3)_3$ with M Na_4EDTA (Priestley, 1963a; see also Section 3.49).

(*iii*) DETERMINATION OF FERRIC CHLORIDE WITH IODINE

Barthel and Schmahl (1965) have titrated ferric chloride in 0·1 N H_2SO_4 with potassium iodide. The results agreed with $\pm 2\%$ with no error whereas the thiosulphate volumetric back-titration method gave results agreeing within $\pm 1\%$ but with a bias of -1.1%.

(*iv*) DETERMINATION OF IRON(III) AS PERIODATE

In a paper describing the determination of mercury by thermometric titration with approximately 0·19 M $NaIO_4$, Tejam and Haldar (1969) also showed that if iron(III) were present it could be titrated quantitatively after the mercury as an endothermic reaction because of the formation with a break point at a 1 : 2 ratio of Fe(III) to IO_6^{5-}. In two tests in which iron(III) was titrated in the presence of mercury, known contents of 93·3 and 18·5 mg were determined as 93·0 and 18·4 mg respectively. The method is selective for iron; silver and lead, if present, will also be titrated quantitatively and thus interfere. Further details of the method are given in Section 3.44.1.

3.36.3 *Properties of the iron cyanide complexes*

(*i*) DETERMINATION OF HEXACYANOFERRATE(II) BY OXIDATION

The determination of hexacyanoferrate(II) by an indicator titration is difficult because of the strong colour of the substance, especially if the solution is concentrated. Mayr and Fisch (1929), using the early titration method, determined 0·7 g of $K_4[Fe(CN)_6]$ in 60 ml of solution by titration with potassium permanganate but reported trouble with heat of dilution because of the need to have a higher concentration of sulphuric acid in the titration solution with the higher concentrations of ferrocyanide. Tyson *et al.* (1961) used ceric oxidation to titrate 0·45 meq of $K_4[Fe(CN)_6]$ in about 50 ml of solution with 0·348 M $Ce(SO_4)_2$. No precision figures were given for this differential titration but the error was stated to be less than $-1·6\%$. Ewing (1961), and Jordan and Ewing (1960) obtained $\Delta H°$ and $\Delta S°$ values of $-9·9 \pm 0·3$ kcal/eq and $+25 \pm 1$ cal eq^{-1} degK^{-1} respectively for this reaction.

(ii) DETERMINATION OF HEXACYANOFERRATE(III) WITH SILVER NITRATE

Silver nitrate has been successfully used as a titrant for the determination of hexacyanoferrate(III), using a burette–thermopile apparatus (Harries, 1966).

(iii) DETERMINATION OF SODIUM PENTACYANONITROSYLFERRATE(III)

Harries (1966) reported that this compound (sodium nitroprusside; $Na_2[Fe(CN)_5NO].2H_2O$) had been titrated with both mercurous nitrate and silver nitrate, using the same apparatus as in (ii) above.

(iv) COMPOSITION OF THE IRON CYANIDE COMPLEXES

The compositions of the iron hexacyanoferrate complexes have been studied by using the early titration method, and the results are summarized in Table 3.50.

Table 3.50 The composition of iron hexacyanoferrate complexes

Titrand	Titrant	Complexes formed	Reference
$K_3[Fe(CN)_6]$ $FeSO_4$	$FeSO_4$ $K_3[Fe(CN)_6]$	$KFe[Fe(CN)_6] \rightarrow Fe_3[Fe(CN)_6]_2$ $Fe_3[Fe(CN)_6]_2$	Bhattacharya and Saxena (1952b)
$K_3[Fe(CN)_6]$ $FeCl_3$	$FeCl_3$ $K_3[Fe(CN)_6]$	$Fe[Fe(CN)_6]$	Bhattacharya and Saxena (1952a)
$K_4[Fe(CN)_6]$ $FeSO_4$	$FeSO_4$ $K_4[Fe(CN)_6]$	$K_2Fe[Fe(CN)_6] \rightarrow Fe_2[Fe(CN)_6]$ $K_2Fe[Fe(CN)_6]$	Saxena and Bhattacharya (1952)
$K_4[Fe(CN)_6]$ $FeCl_3$	$FeCl_3$ $K_4[Fe(CN)_6]$	$KFe[Fe(CN)_6] \rightarrow Fe_4[Fe(CN)_6]_3$ $KFe[Fe(CN)_6]$	Saxena and Bhattacharya (1951) Pâris and Le Chatelier (1934)

The compositions of other metal hexacyanoferrates are discussed under the appropriate sections (*cf.* Cd, Cu, Hg, Pb, Ni, Ag, Tl, Zn).

3.36.4 *Reaction of iron*(III) *with fluorides*

Very sharp inflections were obtained in the thermometric titration curves when 0·82 M NaF was titrated with 1·01 and 2·02 M $FeCl_3$. The inflections were due to $[FeF_6]^{3-}$ ion formation, which was being studied.

If a mixture of iron and aluminium chlorides was titrated they could be distinguished and it was also shown that if the $[FeF_6]^{3-}$ ions were titrated with boric acid the titration curve had an inflection due to the formation of what was thought to be a 'mixed' fluorohydroxoborate BF_3OH^- (Deschamps *et al.*, 1968).

3.36.5 *Formation of the iron(III) complexes of phosphorous acids*

The formation of iron(III) complexes with hypophosphorous acid, phosphoric acid and pyrophosphoric acid have all been studied by using the early titration method and the results are summarized in Table 3.51.

Table 3.51 Formation of iron(III) complexes of phosphorous acids

Solutions used	Complexes formed	Reference
Na_3PO_2 $FeCl_3$	$Fe[Fe(H_2PO_2)_6]$; $FePO_4$; $Fe[Fe(HPO_2)_3]$	Banerjee (1950a)
H_3PO_4 $FeCl_3$	$[Fe(HPO_4)]^+$; $Fe_2(HPO_4)_3$; $[Fe(HPO_4)_2]^-$	Banerjee (1950b)
$Na_4P_2O_7$ $FeCl_3$	$[Fe(P_2O_7)_2]^{5-}$; $[Fe(P_2O_7)]^-$; $Fe_4(P_2O_7)_3$	Banerjee and Mitra (1951)

3.36.6 *Formation of organic acid chelates of iron(III)*

The oxalate, malonate, tartrate and citrate chelates of iron(III) have been studied by titrating ferric nitrate solutions with solutions of the acids. Potassium nitrate was added to the iron solution to increase the ionic strength, and the results obtained are given in Table 3.52.

Table 3.52 Composition of organic acid chelates of iron(III)
(From Gallet and Pâris (1967a, b))

Ferric nitrate	Titrant	Chelates formed Fe^{3+} : *Acid*
2·007 M, pH $<$ 1	Oxalate, 1·352 M, pH 6	1 : 3, 1 : 2
1·005 M, pH 2	Malonate, 3·304 M, pH 5	1 : 1, 1 : 3
0·5002 M, pH 1·5	Tartrate, 1·267 M, pH 11·5	1 : 1·05, 5 : 4, 1 : 1, 1 : 1·5, 1 : 2
0·5013 M, pH 2	Citrate, 1·017 M, pH 10	3 : 2, 1 : 1, 1 : 1·5

The tartrate and citrate chelates of iron(III) were also investigated by Bobtelsky and Jordan (1947), using the early titration method.

3.36.7 *Formation of the phenol complexes of iron(III)*

Using the early titration method, Banerjee and Haldar (1950) found that when a phenol solution was titrated with ferric chloride, the thermometric titration curve showed two inflections corresponding to ratios of Fe^{3+} to phenol of 1 : 6 and 1 : 3. From these results, in conjunction with potentiometric data, it was established that the complex ion $[Fe(OC_6H_5)_6]^{3-}$ and the complex salt $Fe[Fe(OC_6H_5)_6]$ were formed.

Note

For the behaviour of iron salts in an investigation of ligand number see Section 3.51.

3.37 Cobalt

3.37.1 *Determination of cobalt(II) as the* EDTA *chelate*

Na_4EDTA (1 M) has been used as a thermometric titrant for cobalt(II). Jordan and Alleman (1957) obtained a very good precision and an error of $\pm 0.1\%$ and stated that 2 mM was the lowest concentration that could be determined with an error of less than 3%. Priestley (1963a), using an automatic digital titrator, titrated M/60 $CoCl_2$ and also obtained a very good relative standard deviation of $\pm 0.3\%$.

3.37.2 *Determination of cobalt(II) with hydrogen peroxide*

The important technique of using one component to decompose another catalytically with the evolution of heat has been used by Sajó and Sipos to determine traces of cobalt by its catalytic effect on the decomposition of hydrogen peroxide. Two methods have been proposed and used for the analysis of nickel metal. In the first, the sample was dissolved in nitric acid, evaporated to dryness and taken up in hydrochloric acid. Acetic acid (50%) was then added and the whole neutralized with 30% NaOH solution and a known excess added. The presence of acetate was necessary to prevent precipitation of cobalt hydroxide during the reaction. After the temperature of the solution had become stable 30% hydrogen peroxide was injected. Table 3.53 shows the high sensitivity of the method and the constancy of the temperature-rise over three minutes after injection.

Table 3.53 Temperature change of the solution after injection
(Dependence on the cobalt concentrations; from Sajó and Sipos (1967b))

Cobalt concentration	Temperature rise, 10^{-3} degC, *in half-minute intervals*					
$\mu g/ml$	$\frac{1}{2}$	1	$1\frac{1}{2}$	2	$2\frac{1}{2}$	3
0.05	2	2	2	2	2	2
0.1	4	4	5	5	5	5
0.5	63	65	64	65	67	68
1.0	175	170	180	185	190	190

From the results a calibration graph was prepared which is only valid for the specific temperature at which the calibration was made, as obviously the catalytic reaction itself is temperature dependent.

In the second method, the cobalt was used to catalyse the oxidation of tartaric acid by hydrogen peroxide and, provided that the tartaric acid concentration was high enough, the catalysed oxidation rate was sufficiently constant over a long enough period for the thermometric measurements to be made. A differential-burette titration apparatus was used, with two thermistors set in opposition. The sample was dissolved in hydrochloric acid to give a solution containing $0.03-0.3$ mg Co^{2+}, 10 ml of 50% tartaric acid solution were added, the solution was neutralized

Table 3.54 Determination of cobalt in known concentrations
(From Sajó and Sipos (1968a))

Cobalt concentration μM	Cobalt taken μg	Cobalt found μg
0·5	5·9	7·1
1·0	11·8	12·4
2·5	29·4	30·6
5·0	59	62
10·0	118	119
25·0	294	306

with 30% NaOH solution and 20 ml excess were added and the whole was diluted to 200 ml. A blank solution was prepared at the same time, and both solutions were brought to room temperature and transferred to the Dewars. To initiate the reaction, 5 ml of 30% hydrogen peroxide were injected into each Dewar and then the blank Dewar was titrated with 1 mM cobalt until the temperature in both Dewars was the same. The titrant contained cobalt treated in the same way as the sample and was stable for a week. Nickel ions inhibit the reaction slightly but can be compensated for by addition of pure nickel to the blank solution. Table 3·54 gives the results obtained for various known concentrations of cobalt.

Table 3.55 Determination of cobalt in nickel by catalytic thermometry
(Comparison of methods; from Sajó and Sipos (1967b, 1968a))

Sample	Cobalt content, %		
	Photometric method	Catalytic thermometric method 'acetate'	'tartrate'
1	0·009	0·009	0·007
2	0·008	0·007	0·010
3	0·07	0·06	0·08
4	0·08	0·09	0·09
5	0·23	0·20	0·27
6	0·27	0·31	0·29

Both these methods have been used to determine the cobalt content of a number of nickel samples, and the results, compared with those of a photometric method, are given in Table 3.55, and it can be seen that very good agreement was obtained between all three methods.

3.37.3 *Reactions of cobalt with cyanide*

(*i*) FORMATION AND DETERMINATION OF COBALT CYANIDES

In the titration of cobalt(II) with potassium cyanide, using the early titration method, it has been shown that the compounds $Co(CN)_2$, $K_3[Co(CN)_5]$ and $K_4[Co(CN)_6]$ were formed (Mondain–Monval and Pâris, 1934; see also Watt, 1966).

Harries (1966) has satisfactorily titrated cobalticyanides with mercurous nitrate, using a burette–thermopile apparatus.

(*ii*) DETERMINATION OF COBALT(III) AS TETRACYANO- COBALTATE(III)

It has been shown that cobalt(III) can be determined by direct injection with cyanide provided nickel is absent (Rády *et al.*, 1967b). The method is the same as that described in Section 3.38.2(*ii*).

3.37.4 *Composition of the cobalt complexes*

(*i*) HEXACYANOFERRATES

The early titration method has been used to determine the composition of cobalt(II) hexacyanoferrates(II) and (III). The complex hexacyano-ferrates(III) found were $CoK_2[Fe(CN)_6]$ and $Co_4K_4[Fe(CN)_6]_3$ (Pâris and Le Chatelier, 1934), and the complex hexacyanoferrate(II) found was $Co_3[Fe(CN)_6]$ (Gaur and Bhattacharya, 1950a).

(*ii*) PYROPHOSPHATES

The titration of alkali metal pyrophosphate solution with cobalt(II) nitrate, by the early titration method, showed that as well as the normal cobalt pyrophosphate the compounds $[Co(P_2O_7)]^{2-}$ and $[Co(P_2O_7)_2]^{6-}$ were formed.

(*iii*) AMMINES

When 1 N $Co(NO_3)_2$ was titrated with 7·6 N NH_4OH by the early titration method it was found that the complex ammines $[Co(NH_3)_n](NO_3)_2$ were formed, where n had the values 1, 2, 3 or 4. The inflection on the titration graph due to the formation of the $n = 1$ compound, was sharp whereas the others were not (Dutoit and Grobet, 1921).

(*iv*) AQUEOUS COMPLEXES

The heat of formation of the aqueous complexes of cobalt(II) and nickel (II) perchlorates has been studied in n-butanol solution by thermometric titration with water. It was necessary to correct for the heat of mixing of water and n-butanol (Harris and Moore, 1968).

Note

The thermometric titration of cobaltinitrite with silver nitrate failed because the heat of reaction was too small (Harries, 1966).

3.38 Nickel

3.38.1 *Determination of nickel(II) as the* EDTA *chelate*

Priestley (1968a) used an automatic digital titrator and found that a relative standard deviation of $\pm 0.6\%$ could be obtained in eleven replicate titrations of $M/60$ $NiSO_4$. Nickel concentrations (in arbitrary units) between 0·200 and 0·800 were titrated, using a continuous-flow titrator, and the results were within $\pm 2\%$ (Priestley *et al.*, 1965). Jordan and Alleman (1957) titrated 0·01 M Ni^{2+} and obtained a precision of 0·4%, an error of 0·5% and stated that the lowest limit of determination, with an error of 3%, was 1 mmole/l. In all cases M Na_4EDTA was used as the titrant.

3.38.2 *Reaction of nickel(II) with cyanides*

(*i*) FORMATION OF NICKEL CYANIDE COMPLEX

The early titration method was used to study the formation of the tetracyanonickel(II) complex; nickel(II) salts were titrated with potassium cyanide solution and the compound $Ni(CN)_2$ and the complex $K_2[Ni(CN)_4]$ were shown to be formed (Mondain-Monval and Pâris, 1934). The value of the enthalpy of this reaction determined at 10° C using precise apparatus was found to be $\Delta H° = -45.2 \pm 0.6$ kcal/mole (Izatt *et al.*, 1971).

The same reaction was studied by Rasmussen and Nielsen (1963) as a method for the determination of nickel(II). The stability constant of the complex $[Ni(CN)_4]^{2-}$ ion is $\log K = 22$ and the reaction is therefore virtually complete at the equivalence point. The authors used this fact as a basis for their method, and to avoid the precipitation of the intermediate $Ni(CN)_2$ they added ammonia as in the titration of silver (see Section 3.39.2). A high concentration of ammonia (1 M) was required and this caused poor end-points. When ammonium chloride was used in place of ammonia, sharp end-points were obtained even if up to 3 moles of free ammonia were also present, provided that 12 moles of ammonium chloride were present. The results showed a precision of better than $\pm 0.3\%$ and for concentrations greater than 0·01 M Ni^{2+} the disagreement with the electrolytic determination was of the order of 0·3%.

(*ii*) DETERMINATION OF NICKEL(II) AS THE CYANIDE
    COMPLEX

The exothermic reaction of nickel to form the complex cyanide as in (*i*) above has been used to determine nickel in nickel oxides by a direct-injection method (Rády *et al.*, 1967b). The sample, dissolved in dilute sulphuric acid to give a solution containing between 10 and 150 mg of Ni, was diluted with 20% potassium sodium tartrate (Rochelle salt) solution, made alkaline to pH 13·5 with 10% NaOH solution to increase the stability of the complex and injected with potassium cyanide titrant (500 g of KCN + 100 ml of 30% NaOH solution + 400 ml of H_2O). It was found that iron(II) and (III) interfered with the method, but iron(III) did not interfere in the presence of the tartrate.

The method also determined cobalt quantitatively, and as this is a common contaminant of nickel the interference was a serious one. The authors attempted to overcome this by the addition of glyoxime which reacts with both nickel and cobalt ions. The cobalt salt was found to be stable towards potassium cyanide whereas the nickel salt was not, but as its reaction with cyanide had only a small heat of reaction it was not very suitable as a thermometric method. The authors concluded, however, that in many instances it is valuable to know the sum of the nickel and cobalt

Table 3.56 Determination of 94·8 mg of nickel in the presence of other ions
(From Rády *et al.* (1967b))

Other ions present, mg		'Nickel' found	Deviation
Co(II)	Fe(III)	mg	%
—	—	95·0	+0·2
—	—	95·7	+1·0
—	—	94·6	−0·2
—	—	95·2	+0·4
—	—	94·0	−0·8
9·5	—	105·0	+0·7
47·5	—	141·0	−0·9
9·5	10	103·8	−0·5
9·5	10	104·0	−0·3

contents. The relationships between the temperature rise and the nickel and cobalt contents were both linear. The two calibration lines had different slopes, but the difference between them was not sufficient to affect the results provided that the cobalt content was not too large. The results obtained by using the method, given in Table 3.56, show that the sum of the nickel and cobalt is within 1% of the known sum.

Results of the determination of the nickel plus cobalt contents of a number of nickel oxides containing up to 1% of cobalt were found to be well within 2·5% of the sum of their gravimetrically determined contents.

A similar method has been used by Sajó and Sipos (1967a) to determine the nickel content of nickel-plating solutions, but they added sufficient glycerol to the titrant to produce a solution with no heat of dilution when it was injected into a blank solution. The results obtained on a number of plating solutions are given in Table 3.57 and agree well with those from an EDTA volumetric method.

Table 3.57 Determination of nickel in plating solutions
(Comparison of methods; from Sajó and Sipos (1967a))

Sample	Volumetric method g/l	Enthalpimetric method g/l
1	94·3	94·7
2	83·0	81·7
3	64·0	64·6
4	52·7	52·8
5	48·0	47·2
6	37·6	37·4
7	10·1	10·2

3.38.3 *Determination of nickel(II) with* H_2QDT

Beezer and Slawinski (1971) have developed a direct-injection method for the determination of nickel, using quinoxaline-2,3-dithiol (H_2QDT) as the reactant. When 0·6 ml of 1 M $Na_2QDT·2H_2O$ was injected into 25 ml of nickel(II) solution containing 0·5 ml of a pH 10 buffer the following results were obtained, which are compared with thermometric titration results with the same titrant.

Table 3.58 Determination of nickel(II) with H_2QDT
(From Beezer and Slawinski (1971))

Concentration Ni(II)	Enthalpimetric method		Thermometric method	
	No. of replicates	Standard deviation, $\pm\%$	No. of replicates	Standard deviation, $\pm\%$
M/100	6	1·0	3	0·4
M/125	6	1·2	3	1·0
M/167	6	1·4	—	—
M/250	6	1·6	3	1·7

3.38.4 *Reaction of nickel with hexacyanoferrates*

The composition of both the iron(II) and iron(III) cyanide complexes of nickel(II) have been studied by the early titration method. In the titration of nickel sulphate and potassium hexacyanoferrate(II), complexes intermediate between $NiK_2[Fe(CN)_6]$ and $Ni_4K_4[Fe(CN)_6]$ were formed (Pâris and Le Chatelier, 1934). In the nickel sulphate and potassium hexacyanoferrate(III) titration the complex $Ni_3[Fe(CN)_6]_2$ was formed, but in the presence of excess of hexacyanoferrate(III) the complex $KNi_4[Fe(CN)_6]_3$ was also formed (Gaur and Bhattacharya, 1952b).

3.38.5 *Reaction of nickel(II) with bases and amines*

(*i*) REACTION WITH SODIUM HYDROXIDE

Haldar (1948c) investigated the basic sulphates of nickel(II), using the early titration method. The titration of nickel sulphate with sodium hydroxide showed that the compounds $NiSO_4.3NiO$ and $Ni(OH)_2$ were formed, but only the latter compound was formed in the reverse titration.

(*ii*) REACTION WITH AMINES

The enthalpy and entropy of successive steps in the formation of the nickel(II) chelates of ethylenediamine and trimethylenediamine have been determined by a direct-injection method (Poulsen and Bjerrum, 1955).

The formation constant, enthalpy and entropy values, for the association of nickel(II) ions with glycine, phenylalanine and alanine have been determined by titration in aqueous solution at 10, 25 and 40 °C (Anderson *et al.*, 1967), and Holmes and Williams (1967c) determined the thermodynamic properties of the chelation of nickel(II) with 1,2-diaminoethane, 1,3-diaminopropane, 4(5)-aminomethylimidazole, 4(5)-2-aminoethylimidazole, 2-aminomethylpyridine and 2, 2'-aminoethylpyridine.

The reaction of nickel salts with ammonia was investigated by Dutoit and Grobet (1921), but no results are given in the publication.

3.38.6 *Formation of complexes of nickel(II)*

(*i*) HYDROXY-ACID COMPLEXES

Citrate and tartrate solutions have been titrated with nickel(II) by the early titration method and 1 : 1 complexes were shown to be formed. On titration with alkali the citrate complex behaved as a monoprotic acid whereas the tartrate complex was diprotic (Jordan and Ben-Yair, 1957; Ben-Yair and Jordan, 1951).

(*ii*) PYROPHOSPHATE COMPLEXES

The complexes $[Ni(P_2O)]^{2-}$ and $[Ni(P_2O_7)_2]^{6-}$ were shown to be formed when alkali metal pyrophosphates were titrated with nickel nitrate by the early titration method (Haldar, 1950a).

3.38.7 *Formation of the higher nickel oxides*

The titration of strongly alkaline persulphate with M $Ni(NO_3)_2$, using the early titration method, showed that the hydrated forms of NiO_2, Ni_2O_3, Ni_3O_4 and possibly Ni_5O_6 were formed (Glemser and Einerhand, 1950).

Note

For the formation of the aqueous complexes of nickel(II) perchlorate see Section 3.37.7 and for the determination of small amounts of nickel by a catalytic method see Section 3.52.

3.39 Silver

3.39.1 *Determination of silver as* AgCl

This precipitation reaction has been studied extensively as a quantitative thermometric method for silver. These studies are summarized in Table 3.59 and are discussed further in Section 3.27.1(*i*) as the less successful reverse titration.

Table 3.59 Precision and accuracy of thermometric titrations of silver with halides

Titrant			Titrand	Number	Precision	Error	Ref.
Type	M	ml	M	of tests	±%	%	
HCl	1·013	75	0·9960	4	0·1	+0·1	(a)
HCl	1·013	75	0·4899	4	0·2	+0·1	(a)
HCl	1·013	75	0·1989	3	0·7	+1·9	(a)
HCl	0·9698	75	1·003	5	0·4	−0·2	(a)
HCl	0·9698	75	0·2987	10	0·5	0·0	(a)
HCl	1·032	60	5·000	5	0·31	0·0	(b)
HCl	1·008	60	5·000	8	0·25	−0·4	(b)
HCl	0·115	60	0·500	3	0·77	0·0	(b)
HCl	1·000	20	1·000	11	0·6	—	(c)
HCl	0·200	200	10·0	—	0·2	—	(d)
HCl	1·000	30	0·500	10	0·5	—	(e)
NaCl	0·050	80	0·015	—	0·2	—	(f)
NaCl	0·200	—	—	—	—	—	(g)
NaCl	0·200	200	10·0	—	0·2	—	(d)
BaCl$_2$	0·200	200	10·0	—	0·2	—	(d)
LaCl$_3$	0·200	200	10·0	—	0·2	—	(d)

(a)   Tyson *et al.* (1961)—differential titration
(b)   Linde *et al.* (1953)—simple automatic titration
(c)   Priestley *et al.* (1963)—automatic titration
(d)   Barthel *et al.* (1968)—normal titration
(e)   Priestley (1963a)—automatic digital titration
(f)   Jahr *et al.* (1968)—normal titration
(g)   Priestley *et al.* (1968)—continuous flow titration

It was reported (*a*, in Table 3.59) that the titration of silver nitrate with hydrochloric acid was especially accurate except when very dilute solutions of about 0·004 N were titrated and (*c*, in Table 3.59) that the heat produced was greater than that from the neutralization of the same quantity of hydrochloric acid.

3.39.2 *Heat of precipitation of silver halides and silver chromate*

Table 3.60 gives the standard heat of precipitation of the silver halides that have been determined by thermometric titration together with values obtained by other methods for comparison.

**Table 3.60 Heat of precipitation of silver halides:
—$\Delta H°$ kcal/mole in water at 25°C**

AgCl	AgBr	AgI	Error, %	Reference
15·70	19·90	26·94	±0·3	Ewing and Mazac (1966)
15·7	20·1	26·90*	±0·2	Barthel *et al.* (1968)
15·70	—	—	±0·2	Tyson *et al.* (1961)
16·0	—	—	±0·2	Jahr *et al.* (1968)
15·65	20·19	26·85	—	Literature values

* 21·6 in methanolic solution

In their investigation into thermometric titrimetry at high temperature in a eutectic of fused alkali nitrates (43·1 mole % $LiNO_3$ + 56·9 mole % KNO_3) Jordan and co-workers (1958; 1959b; 1960a, b, c) determined the heats, free energies and entropies of the formation of the silver halides and chromate by titration of the respective potassium salts (0·1–20 mm) with a concentrated silver nitrate solution in the eutectic; their results are summarized in Table 3.61. The authors showed that stoichiometric com-

**Table 3.61 Heats and entropies of precipitation of
silver salts at high temperature**
(Results from Jordan (1960); Jordan *et al.* (1960a))

Compound	Temperature °C	$-\Delta H°$ kcal/mole	$-\Delta S°$ cal mole^{-1} degK^{-1}	$-\Delta G°$ kcal/mole
AgCl	158±2	18·9±0·3	—	14·9±0·2
	168±1	18·9±0·9	9·3±2	—
	318±2	19·6±0·6	11·7	—
AgBr	168±1	26·1±0·7	19·5	—
	318±1	26·5±1	17·4	—
AgI	168±1	32·1±1	18·3	—
	318±2	35·1±1	22·0	—
Ag_2CrO_4	168±1	16·7±0·5	7·4±2	15·7±0·4
	250±2	16·9±0·4	11·7±1	—

pounds were formed, though it was found that at low concentrations the end-point was rounded because of incomplete precipitation. However, from this rounding the solubility product for silver chloride could be calculated as 3×10^{-8} mole^2 kg^{-2} under these particular conditions. The value of —18·9 kcal/mole for the heat of precipitation of AgCl obtained from the titration curve agrees reasonably well with the value of —18·3 kcal/mole obtained by Flengas and Rideal (1956) from solubility measurements in KNO_3–$NaNO_3$ eutectic at 250–350°C.

When the procedure was used in the quantitative determination of potassium chloride (Jordan *et al.*, 1959a) it was found that between 0·0939 and 1·985 mmole of the salt could be determined with a relative standard deviation of $\pm 1·5\%$.

3.39.3 *Reaction of silver with cyanides*

(*i*) DETERMINATION OF SILVER AS AgCN

The thermometric titration curve obtained when silver is titrated with potassium cyanide shows a very blurred inflection because of the slow precipitation of silver cyanide. No inflection is obtained when the precipitate dissolves to form the dicyanoargentate(I) ion. This difficulty was overcome by Rasmussen and Nielsen (1963) who titrated after adding enough ammonia to just dissolve the silver oxide formed. A sharp end-point was obtained on the formation of the dicyanoargentate(I) ion. Approximately 3 mmole of silver nitrate were determined with a deviation of less than $\pm 0·7\%$ (mean $\pm 0·4\%$). In one series of tests the normality [*sic*] of the cyanide titrant was found to be $1·570 \pm 0·005$ eq/l compared to $1·575 \pm 0·001$ eq/l found by using an iodide indicator titration (Liebig-Denigès titration). The disagreement between the two methods is thus less than $0·3\%$.

(*ii*) PROPERTIES OF THE SILVER-CYANIDE SYSTEM

The thermodynamic properties, log K, $\Delta H°$ and $\Delta S°$ have been determined for the system $[Ag(CN)_2]^- + CN^- \rightarrow [Ag(CN)_3]^{2-}$ and the properties of $\Delta H°$ and $\Delta S°$ for the system $Ag^+ + 2CN^- \rightarrow [Ag(CN)_2]^-$ (Izatt *et al.*, 1967).

(*iii*) REACTION OF SILVER WITH HEXACYANOFERRATE(II)

Harries (1966), using a burette–thermopile apparatus, titrated potassium hexacyanoferrate(II) with silver nitrate and found that the heterometallic compounds $K[Fe(CN)_6Ag_3]$ and $Ag[Fe(CN)_6Ag_3]$ were formed successively. Similar results were obtained by Pâris and Le Chatelier (1934), using the early titration method.

3.39.4 *Determination of silver as the* EDTA *chelate*

Automatic thermometric titrators have been used to determine silver with M Na_4EDTA. Using an automatic digital titrator, Priestley (1963a) titrated M/60 $AgNO_3$ and obtained a relative standard deviation of $\pm 2·0\%$ with four replicates. He and his co-workers (1963) titrated 1 mmole of Ag^+ in 200 ml of solution and obtained a T value of 0·17 degC/mmole. The titration curve showed an exothermic stage due to the formation of $Na[Ag_3Y]$ which gave the main temperature rise used in the analysis, but it was also followed by an endothermic stage due to $Na_2[Ag_2Y]$ and after that another exothermic stage due to $Na_3[AgY]$ (where Y = EDTA).

3.39.5 *Determination of silver as periodate*

In a paper describing the determination of mercury by thermometric titration with approximately 0·19 M $NaIO_4$ Tejam and Haldar (1969) also showed that if silver were present it could be titrated quantitatively after the mercury as an endothermic reaction with a break point at 2 : 1 ratio of Ag^+ to IO_6^{5-}. In two titrations in the presence of mercury, known contents of 222·9 and 21·6 mg were determined as 223·4 and 21·4 mg respectively. The method is selective for silver, lead and iron, if present, being also titrated quantitatively. Further details of the method are given in Section 3.44.1.

3.39.6 *Determination of silver as carbonate*

Priestley (1963a) has titrated M/60 $AgNO_3$, using an automatic digital titrator and obtained with eight replicate titrations a relative standard deviation of $\pm 1 \cdot 1 \%$.

3.39.7 *Composition of silver complexes*

(*i*) HALIDE COMPLEXES

Halide solutions with a concentration of 10–20 mM have been titrated with 0·1 M $AgNO_3$ in dimethylsulphoxide solution, and it was shown that the species $[AgCl_2]^-$ and $[AgBr_2]^-$ were formed. However, with iodide ions an intermediate between $[AgI_2]^-$ and $[Ag_3I_4]^-$, probably $[Ag_5I_7]^{2-}$ was formed (Jambon and Merlin, 1971).

(*ii*) THIOSULPHATE COMPLEXES

Gaur and Bhadrauer (1959), using the early titration method with 0·2 and 0·1 M $AgNO_3$ as titrant, found that the following compounds were formed successively: $Na_3Ag(S_2O_3)_2$, $NaAg(S_2O_3)$, $Na_5Ag_3(S_2O_3)_4$ and, finally, $Ag_2(S_2O_3)$. With sodium thiosulphate as the titrant only the compounds $NaAg(S_2O_3)$ and $Ag_2(S_2O_3)$ were formed.

Table 3.62 Thermodynamic data for the silver-pyridine interaction at 25 °C

Reaction	$\log \beta_i$	$-\Delta H°$ kcal/mole	$-\Delta S°$ cal mole^{-1} degK^{-1}	Reference
$Ag^+ + Py \rightleftharpoons [AgPy]^+$	2·05 ± 0·05	4·6 ± 0·2	6·0 ± 0·8	Izatt *et al.*, 1968
	—	4·77	5·7	Becker *et al.*, 1963b
$Ag^+ + 2Py \rightleftharpoons [AgPy_2]^+$	4·10 ± 0·05	11·23 ± 0·11	18·8 ± 0·4	Izatt *et al.*, 1968
	—	11·53	19·5	Becker *et al.*, 1963b

(*iii*) PYRIDINE COMPLEXES

The titration of 200 ml of 0·015 M $AgNO_3$ with 0·669 M pyridine (Py) and also with other concentrations has shown that the interaction is an equilibrium, $Ag^+ + Py \rightleftharpoons [AgPy]^+ + Py \rightleftharpoons [AgPy_2]^+$ (Becker *et al.*, 1963b). The same system has been studied more thoroughly by Izatt and co-workers (1968) who used approximately 1 M $AgNO_3$ and pyridine (each as titrand and titrant) and a computer to evaluate the thermodynamic data from the titrations. The results are shown in Table 3.62 (see also Christensen, Rytting and Izatt, 1969).

Note

For the determination of small amounts of silver by a catalytic endpoint method see Section 3.52. It should be possible to use tetraphenylborate as a titrant in the determination of silver (see Section 3.3.2).

3.40 Cadmium

3.40.1 *Determination of cadmium as the EDTA chelate*

Cadmium has an even higher heat of chelation with EDTA than has copper, so precise thermometric titrations of cadmium can be obtained. Jordan and Alleman (1957) titrated 0·01 M Cd^{2+} with M Na_4EDTA and obtained a precision of 0·4% and an error of 0·4%. They quoted a lowest level of determination of 1 mmole/1, with an error of 3%. A relative standard deviation of $\pm0\cdot5\%$ was obtained in eleven replicate titrations of M/60 $CdSO_4$ with M Na_4EDTA, using an automatic digital titrator (Priestley, 1963a).

3.40.2 *Reaction of cadmium with alkalis*

(*i*) FORMATION OF BASIC CADMIUM SALTS

Haldar (1946a) titrated cadmium sulphate with sodium hydroxide, using the early titration method. In all the titrations the final product shown to be formed was $CdSO_4.3Cd(OH)_2$, but with concentrated reagents the intermediate compound $CdSO_4.Cd(OH)_2$ was formed and with cadmium sulphate as titrant the intermediate was $Cd(OH)_2$. Ben-Yair (1961), titrating cadmium nitrate with sodium hydroxide, reported the formation of the compound $CdNO_3.OH$.

(*ii*) DETERMINATION OF CADMIUM

Ben-Yair (1957) reported the determination of cadmium by thermometric titration with ammonia or sodium hydroxide, using the early titration method.

(*iii*) CITRATE AND TARTRATE CHELATES OF CADMIUM(II)

Partially alcoholic solutions of the citrate and tartrate chelates of cadmium have been titrated with M NaOH (or NH_4OH if the NaOH gave insoluble salts) by the early titration method (Jordan and Ben-Yair, 1957). Both the tartrate chelate ($T = 0\cdot40$ degC/mmole) and the citrate chelate ($T = 0\cdot47$ degC/mmole) were found to be monoprotic acids, unlike copper and zinc tartrates, which were found to be diprotic acids. This titration has been used as the basis of a method for the determination of citrates and tartrates (see Section 4.5). The same authors also studied the composition of these hydroxy-acid compounds and showed them to be $1:1$ chelates (Ben-Yair and Jordan, 1951).

3.40.3 *Composition of cadmium(II) complexes*

All the following complexes of cadmium have been studied, the early titration method being used except where stated.

(*i*) CADMIUM HEXACYANOFERRATE(II) COMPLEXES

The composition of cadmium hexacyanoferrate(II) complexes has been studied by Pâris and Le Chatelier (1934) who reported that the compounds $K_2Cd[Fe(CN)_6]$ and $K_4Cd_4[Fe(CN)_6]_3$ were formed. The same compounds were found by Bhattacharya and Gaur (1948) but they also reported the formation of intermediates between these two compounds. The composition of cadmium hexacyanoferrate(III) has also been studied by Gaur and Bhattacharya (1949) who used aqueous and partially alcoholic solutions for the titration. The compound $KCd_{10}[Fe(CN)_6]_7$ was formed when either cadmium sulphate or potassium hexacyanoferrate(III) was used as the titrant.

(*ii*) CADMIUM CYANIDE COMPLEXES

A very precise apparatus has been used in the determination of the heat of formation at 10, 25 and 40 °C of cadmium cyanide complexes, $Cd(CN)_n$, where $n = 1$, 2, or 4. In the determination the metal chlorate solution was titrated with cyanide to a $CN:Cd^{2+}$ ratio of $2:1$. A second solution, made up to the same ratio of $2:1$, was also titrated to a ratio of greater than $4:1$. Perchloric acid was added to the titrand to prevent hydrolysis (Izatt *et al.*, 1971).

(*iii*) CADMIUM HALIDE COMPLEXES

Chatterji (1958b) found that when cadmium sulphate was titrated with 3 M KI the complex ion $[CdI_4]^{2-}$, and possibly $[CdI_5]^{3-}$, was formed. These complex ions and also $[CdI_3]^-$, $[CdI_6]^{4-}$ and $[CdI_7]^{5-}$ were said to be found in a similar study by Haldar (1946d). In the same work Haldar also titrated cadmium iodide and showed that with 5 M KI as titrant the complex ions $[CdI_5]^{3-}$ and $[CdI_3]^-$ were formed but that $[CdI_4]^{2-}$ was formed in the place of the latter complex when more dilute titrants were used. Becker and Lüschow (1964) also showed that a $[CdI_4]^{2-}$ complex was formed.

Gerding (1966a, b), using modern apparatus, determined the free energy, enthalpy and entropy changes in the stepwise formation of cadmium complexes in the reaction $[CdX_{i-1}]^{3-i} + X^- = [CdX_i]^{2-i}$ where X^- was F^-, Cl^-, CN^-, N_3^-, SCN^-, or NO_2^-. All the results, except those for F^-, were obtained in a 3 M $NaClO_4$ medium to reduce heat of mixing from the 1–3 M titrants. Metzger *et al.* (1967) studied a molten $KCl–CdCl_2$ system, using a calorimeter in which one of the salts could be added incrementally to the melt of the other. A $1:1$ complex was formed, having an instability constant of 0·32 at 600 or 780 °C.

(*iv*) CADMIUM THIOSULPHATE COMPLEXES

The titration of cadmium chloride with sodium thiosulphate or *vice versa* in aqueous or partially alcoholic solution showed that only the compound CdS_2O_3 was formed (Bhadrauer and Gaur, 1959).

(*v*) CADMIUM THIOUREA COMPLEXES

Siddhanta (1947) titrated 40 ml of 0·2146 M $CdCl_2$ with 2·022 M $SC(NH_2)_2$. The thermometric curve showed three inflections corresponding approximately to the complexes $CdCl_2.2SC(NH_2)_2$, $CdCl_2.4SC(NH_2)_2$ and $CdCl_2.6SC(NH_2)_2$. The inflections were not stoichiometric for these compounds as it was found that dissociation occurred and a 6–7% excess of titrant was required to complete the reaction.

3.40.4 *Ion-association in the cadmium–sulphate system*

In an investigation into the possible formation of anionic cadmium sulphate complexes of the type $[Cd(SO_4)_2]^{2-}$ in concentrated solution, Becker and Grundmann (1969) added sodium perchlorate to the titrand (0·08 or 0·16 M $Cd(ClO_4)_2$) to obviate the large heat of dilution of the 2 M Li_2SO_4 titrant. No anionic complexes were formed but the following results were obtained for the 1 : 1 ion-association complex: $K = 4·3 \pm 0·2$; $\Delta H° = 1·90 \pm 0·10$ kcal/mole; $\Delta S° = 9·2 \pm 0·4$ cal mole^{-1} degK^{-1}.

Note

The titration of cadmium acetate is discussed in Section 4.1.4(*ii*). For the behaviour of cadmium salts in an investigation of ligand number by thermometric titrimetry see Section 3.51 and for the determination of small amounts of cadmium by a catalytic method see Section 3.52.

3.41 Zirconium

The free acidity in zirconyl salts has been determined thermometrically and is discussed in Section 3.48.1(*viii*).

3.42 Molybdenum

3.42.1 *Reaction of molybdenum with hydrogen peroxide*

(*i*) FORMATION OF MOLYBDENUM SALTS

The thermometric titration of 75 ml of hydrogen peroxide solution in 5 N H_2SO_4 with 0·506 M MoO_3 (as paramolybdate) (Rivenq 1945a; early titration method) gave a curve with a sharp inflection showing that the reaction could be represented by $MoO_3 + 2H_2O_2 \rightarrow MoO(O_2)_2 + 2H_2O$.

(*ii*) DETERMINATION OF MOLYBDENUM AS THE PEROXY SALT

The exothermic reaction above has been used as a basis for the determination of molybdenum in alloys containing between 0·1 and 100% Mo (Sajó *et al.*, 1967). The sample of filings (1 g) was dissolved in hydrochloric acid and oxidized with nitric acid. The solution was evaporated to dryness and taken up in hydrochloric acid. Sulphuric acid was then added and the solution diluted to 200 ml, and if titanium were present hydrofluoric acid was added to complex it and so prevent it from reacting with the titrant. The solution was then injected with a 30% solution of hydrogen peroxide. If hydrofluoric acid had been added precautions were taken to prevent attack on the apparatus. Table 3.63 gives the results obtained by using the method compared with those of a very lengthy gravimetric oxime method involving prior separation on an ion-exchange resin column.

Table 3.63 Determination of molybdenum in alloys
(Comparison of methods; from Sajó *et al.* (1967))

Sample	Gravimetric method, %	Enthalpimetric method, %
1	5·28	5·20
2	3·04	3·10
3	4·25	4·30
4	5·94	5·90

The authors stated that cobalt, nickel, aluminium, manganese and chromium did not interfere with the method, but presumably chromium would interfere if it were present in the Cr(VI) state as it then reacts with hydrogen peroxide (see Section 3.34.3).

3.42.2 *Reaction of molybdenum with boron trifluoride*

Richards and Woolf (1969) studied the reaction of boron trifluoride containing bromine with molybdenum(VI) ions and showed that molybdenum hexafluoride was formed.

3.43 Palladium

The only reference to a thermometric method for the determination of palladium is the catalytic one described in Section 3.52.

3.44 Mercury

3.44.1 *Determination of mercury(II) with periodate*

Tejam and Haldar (1969) have shown that in the titration of mercury(II) nitrate with either periodic acid, sodium metaperiodate or with potassium dimesoperiodate an inflection occurs in the titration curve at a ratio of $Hg:IO_6^{5-}$ equal to $5:2$. The inflection was reproducible and sharp, particularly when sodium metaperiodate was used, which was preferred as titrant because it was also easy to prepare.

In this important thermometric method, as in the gravimetric method (Vogel, 1964), there was no interference from the following ions: Al^{3+}, AsO_4^{3-}, Ba^{2+}, BO_3^{3-}, Be^{2+}, Cd^{2+}, Ca^{2+}, Co^{2+}, CrO_4^{2-}, Cu^{2+}, Mg^{2+},

Mn^{2+}, MnO_4^-, Ni^{2+}, PO_4^{3-}, SO_4^{2-}, Sr^{2+}, UO_2^{2+}, Zn^{2+}. If sufficient nitric acid were present interference from Bi^{3+} and Sb^{3+} ions was suppressed, but though 1 M HNO_3 had no effect on the reaction, higher concentrations reduced the sharpness of the end-point. Nitrites and sulphites were oxidized with permanganate, and cyanide destroyed by boiling the solution containing nitric acid before the titration. If the ions Ag^+, Pb^{2+} and Fe^{3+} were present they did not interfere as they do in the gravimetric method, but were titrated quantitatively after the mercury as an endothermic reaction so the end-point between these metals and mercury was sharp. The final end-point was also sharp as the titrant had a positive heat of dilution. These metal ions can therefore be determined at the same time but cannot be distinguished from one another in the titration. It follows that silver nitrate can be used to remove interfering chloride.

Table 3.64 gives the results obtained when 20 ml of mercury(II) nitrate solution, acidified with nitric acid, were titrated with the three titrants at approximately 0·2 M concentration, both alone and in the presence of other ions.

Table 3.64 Titration of mercury(II) with periodates
(From Tejam and Haldar (1969))

Titrant (about 2 M)	Mercury Present	Found	Other ions present	Error %
Periodic acid	20·0	20·0	—	0·0
H_5IO_6	386·8	386·6	—	$<-0·1$
Potassium dimesoperiodate	20·0	19·9	—	−0·5
$K_4I_2O_9$	421·3	422·0	—	−0·2
Sodium metaperiodate	26·1	26·2	—	+0·4
$NaIO_4$	386·8	387·3	—	+0·1
	21·6	21·5	Ag^{1+}	−0·5
	491·1	490·0	Ag^{1+}	−0·2
	20·4	20·3	Pb^{2+}	−0·5
	491·1	491·9	Pb^{2+}	+0·2
	20·4	20·3	Fe^{3+}	−0·5
	483·6	485·8	Fe^{3+}	+0·5

3.44.2 *Reaction of mercury(II) with halides*

(*i*) DETERMINATION OF MERCURY(II) AS $HgCl_2$

Mercury(II) nitrate has been titrated with hydrochloric acid in a differential titration apparatus (Tyson *et al.*, 1961), and although no quantitative measurements were recorded the authors noted that the titration curve had two inflections. The first inflection caused by the formation of $HgCl^+$ showed a greater heat of reaction than the second inflection caused by $HgCl_2$. Mercury(II) nitrate (10 meq in 60 ml) has been titrated with 1 M

HCl, in a simple automatic titrator. Five replicates had a precision of $\pm 0.35\%$ and an error of $+0.1\%$ (Linde *et al.*, 1953).

(*ii*) MERCURY(II) HALIDE COMPLEXES

Arnek (1965) used a thermometric titration to study the systems Hg(II)–Cl and Hg(II)–Br at 25 °C and determined the enthalpy change and equilibrium constant of the reactions.

It was shown, using the early titration method, that when potassium iodide was titrated with mercury(II) chloride the compounds HgI_2 and K_2HgI_4 were formed (Mondain-Monval and Pâris, 1934). Barthel and co-workers (1968), investigating the reaction as a quantitative method, found that in the titration with potassium iodide, the HgI_2 end-point was satisfactory whereas the $HgI_4{}^{2-}$ one was not.

(*iii*) MERCURY(II)–POTASSIUM BROMIDE COMPLEXES

The reaction of mercury(II) bromide with potassium bromide has been studied by thermometric titration, and the complex ions $[HgBr_3]^-$ and $[HgBr_4]^{2-}$ were shown to be formed (Becker *et al.*, 1963b). The thermodynamic data obtained from the titration curves are given in Table 3.65 (see also Bjorkman and Sillén, 1963).

Table 3.65 Thermodynamic data for the $HgBr_2 + 2Br^- \rightleftharpoons [HgBr_4]^{2-}$ reaction in water at 25 °C
(From Becker *et al.* (1963b))

Reaction	$-\Delta G^\circ$ kcal/mole	$-\Delta H^\circ$ kcal/mole	ΔS° cal mole^{-1} degK^{-1}
$HgBr_2 + Br^- \rightleftharpoons$ $[HgBr_3]^-$	3·09	2·08	$+3\cdot4$
$HgBr_2 + 2Br^- \rightleftharpoons$ $[HgBr_4]^{2-}$	4·97	5·54	$-1\cdot9$

3.44.3 *Determination of mercury*(II) *as* $Hg(CN)_2$

The heat of conversion of mercury(II) chloride in solution into a precipitate of mercury(II) cyanide and the subsequent heat of dissolution of it to form the tetracyanomercurate(II) ion are respectively 33·2 and 12·4 kcal/mole (Landolt-Börnstein, 1923). Recent work in which a precise apparatus was used to determine the heats of formation at 10, 25 and 40 °C of mercury(II) cyanide complexes, $Hg(CN)_n$, where $n = 1, 2, 3$ or 4, gave the values for the same reactions at 25 °C as 46·8 and 13·1 kcal/mole respectively (Izatt *et al.*, 1971). This indicates that the thermometric curve should show two inflections and this was verified, using the early titration method, when the compounds $Hg(CN)_2$ and $[Hg(CN)_4]^{2-}$ were shown to be formed when mercury(II) chloride was titrated with potassium cyanide (Mondain-Monval and Pâris, 1938b).

Dissolution of the precipitated cyanide is a slow process and because of this the end-point showing the formation of the tetracyanomercurate(II) ion is curved, even in the presence of excess of potassium chloride. The cyanide precipitation end-point is sharp and was investigated as a possible method for the determination of mercury (Rasmussen and Nielsen, 1963). Between 1·61 and 2·45 mmole of mercury(II) chloride in about 100 ml of solution were titrated with 1·55 M KCN. The results of six determinations had a deviation of $+1·2\%$ and from this it was concluded that this method was not satisfactory.

3.44.4 *Determination of mercury(I) and (II) as oxalates*

The reaction of mercury with oxalate is not a recognized method for its determination but Mayr and Fisch (1929), using the early titration method, showed that it was possible to titrate both mercury(II) and mercury(I) nitrates and also to distinguish between them in admixture. With 5 M $(NH_4)_2C_2O_4$ as titrant the end-point of the titration of each ion was clearly marked. The results of the titration of mercury(II) ions both alone and in the presence of mercurous ions are correct to well within $0·1\%$ but the titration of mercury(I) ions in the presence of mercury(II) ions is less accurate at low values, as can be seen from Table 3.66.

Table 3.66 Titration of mercury(I) and (II) with oxalate
(From Mayr and Fisch (1929))

Sample	*Mercury*(II), g		*Mercury*(I), g	
	Present	*Found*	*Present*	*Found*
1	0·8886	0·8882	—	—
2	0·8886	0·8886	0·0855	0·0967
3	0·8886	0·8886	0·3481	0·3442

3.44.5 *Determination of mercury(II) as* $HgNH_2Cl$

Ben-Yair (1957, 1961) has shown that when mercury(II) chloride is titrated with ammonia, the complex $HgNH_2Cl$ is formed. He used this as a method for the determination of mercuric salts and although detailed results are not available, claimed that it gave better results than the potentiometric method.

3.44.6 *Formation of mercury(II) complexes*

(*i*) HEXACYANOFERRATE(II) COMPLEXES

Harries (1966), using a burette–thermopile apparatus, titrated potassium hexacyanoferrate(II) with mercury(II) nitrate. The titration curve showed two inflections, the intermediate one corresponding to the complex $K_2[Fe(CN)_6Hg_2]$ which was unstable and could not be isolated and a final inflection corresponding to the complex $Hg_2[Fe(CN)_6Hg_2]$ which was stable and could be isolated.

(ii) 2-Aminoethanol complex

The equilibrium constant and other thermodynamic data for the formation of the extremely stable mercury(II)–2-aminoethanol complex has been determined by Eatough (1970). The interaction was studied by the titration of two different solutions containing the complex perchlorate and perchloric acid, with perchloric acid. A value of $8 \cdot 56 \pm 0 \cdot 07$ was obtained for log K and $-10 \cdot 2 \pm 0 \cdot 5$ kcal/mole for $\Delta H°$.

Note

The titration of mercury(II) acetate is discussed in Section 4.1.5(*iv*). For the behaviour of mercury(II) salts in an investigation of ligand number see Section 3.51 and for the determination of small amounts of mercury by a catalytic method see Section 3.52.

3.45 Tungsten

3.45.1 *Formation of the tungsten peroxy salts*

Rivenq (1945b), using the early titration method, titrated 50 ml of hydrogen peroxide with $0 \cdot 301$ M Na_2WO_4 and a rounded end-point indicated that the reaction could be represented by the equation $WO_3 + 2H_2O_2 \rightarrow WO(O_2)_2 + 2H_2O$ ($T = 0 \cdot 37$ degC/mmole).

3.45.2 *Reactions of the isopoly- and heteropolytungstates*

The decomposition of the isopoly- and heteropolytungstates by sodium hydroxide has been studied by thermometric titrimetry using $1 \cdot 6$ M NaOH as the titrant (Spitsyn and Kosmodem'yanskaya, 1965; 1966a, b; 1968). The compounds $Na_2W_4O_{13}$, $Na_{10}W_{12}O_{41}$, $H_4SiW_{12}O_{40}$, $Na_3PW_{12}O_{40}$ and $H_3PW_{12}O_{40}$ were studied and the mode of their decomposition to Na_2WO_4 was ascertained.

3.46 Lanthanides

The thermodynamic properties of the lanthanide complexation reactions have been determined thermometrically. The complexes studied were as follows:

Monosulphates	Fay and Purdie, 1969
Mono- and di-sulphates	De Carvalho and Choppin, 1967
Acetates	Grenthe, 1964; Grenthe and Williams, 1967
α-Hydroxy-carboxylates	Friedman, 1966
Glycollates and thioglycollates	Grenthe and Williams, 1967

Further studies have been concerned with cerium.

3.46.1 *Determination of cerium(IV) by reduction with iron(II)*

The enthalpy and entropy of the reduction of Ce(IV) with Fe(II) have been determined at 25 °C by enthalpimetric titration in sulphuric acid medium. The results are summarized in Table 3.67.

Table 3.67 The enthalpy and entropy of the reduction of Ce(IV) with Fe(II)

H_2SO_4 N	$-\Delta H°$ kcal/mole	$-\Delta S°$ cal mole^{-1} degK^{-1}	*Reference*
0·5	24·1 ± 0·6	22·0 ± 2	Jordan and Ewing (1960)
0·5	24·2 ± 0·6	23·0 ± 2	Ewing (1961)
0·2	24·4 ± 0·5	—	Barthel and Schmahl (1965)
0·5	23·0 ± 0·5	—	Barthel and Schmahl (1965)
1·0	22·5 ± 0·5	—	Barthel and Schmahl (1965)

The high heat of reaction indicates that precise and sensitive thermometric cerium(IV)–iron(II) titrations are possible (see also Section 3.36.1).

Barthel and Schmahl (1965) have titrated 0·01 M $Ce(SO_4)_2$ in 200 ml of 0·2–1·0 N sulphuric acid. With 0·2 M $Fe(NH)_2(SO_4)_2 \cdot 6H_2O$ as titrant the results obtained were precise to within ±0·3% compared to ±0·5% when a ferroin indicator was used; the latter method also showed a negative bias. ($T = 0·23$ degC/mmole.) In a similar titration, using an incremental titration method (Brown *et al.*, 1969), nine replicate titrations of 0·025 M Ce^{4+} with 0·5 M Fe^{2+} gave no bias and a relative standard deviation of 0·21%.

Recent work, using a high-precision incremental titration procedure, has shown that 0·025 M Ce(IV) can be titrated with 0·05 M Fe^{2-}, and in nine replicate titrations a standard deviation of ±0·21% was obtained (Brown *et al.*, 1969). Guillot (1970), using an automated direct-injection method, injected 25 ml of 0·2 M $Fe(NH_4)_2(SO_4)_2 \cdot 6H_2O$ solution in 0·5 M H_2SO_4 into 2 ml of 0·2–1·0 M Ce(IV) and obtained results correct to within 0·4%.

3.46.2 *Determination of cerium(IV) with thiosulphate*

Priestley (1963a), using an automatic digital titrator, showed that M/30 $Ce(SO_4)_2$ in dilute sulphuric acid solution could be titrated with 2 M $Na_2S_2O_3$, and in eight replicate titrations obtained a relative standard deviation of ±0·6%.

3.47 Actinides; thorium and uranium

3.47.1 *Thorium complexes*

Mixed-ligand chelates composed of two different multidentate ligands linked to a central thorium(IV) ion have been prepared in aqueous solution and the thermodynamic data for their heats of formation have been studied by titration of the complex with 0·5 M NaOH. A 4 mM ligand and/or metal ion concentration was used, and the solutions examined were maintained at constant ionic strength with potassium nitrate. The primary ligands used were ethylenediaminetetra-acetate and 1, 2-diamino-cyclohexane-*N,N,N′,N′*-tetra-acetate, and the secondary ligands were catechol, chromotropic acid, 8-hydroxyquinoline-5-sulphonic acid, imino-diacetic acid, salicylic acid, 5-sulphosalicylic acid and Tiron (Kugler and Carey, 1970).

The formation of the thorium(IV) oxalate complexes has been studied, using the early titration method (Bose and Chowdhury, 1955). Potassium oxalate solution was titrated with thorium nitrate solution and the complex species $[Th(C_2O_4)_4]^{4-}$, $[Th(C_2O_4)_3]^{2-}$ and $[Th(C_2O_4)_2]$ were shown to be formed.

3.47.2 *Determination of uranium(IV) by dichromate oxidation*

Miller and Thomason (1959b) have determined 11 mg of uranium tetrachloride in 5 ml of acid solution by titration with 1 N $K_2Cr_2O_7$ in 1 N H_2SO_4 with a relative standard error of $\pm1\%$ and no bias. They found that zirconium(IV), aluminium(III), sulphate and chloride did not interfere. Fluoride did not interfere provided that sufficient aluminium was present to complex it; for example, zirconium–uranium fluorides were dissolved in an aluminium chloride–sulphuric acid solution. The method was stated to be superior to volumetric redox methods involving electrodes that were attacked by the ions present in the samples.

3.47.3 *Reaction of uranyl salts with carbonate and hexacyanoferrate(II)*

Indian workers used the early titration method to determine the composition of uranyl carbonate and hexacyanoferrate(II). In the titration of sodium or potassium carbonate with 0·4 M $UO_2(NO_3)_2$ the compounds $R_4UO_2(CO_3)_3$ and $R_2UO_2(CO_3)_2$ were formed, where R was the alkali metal. The same compounds were formed in the reverse titration (Haldar, 1947, 1948a).

In the titration of potassium hexacyanoferrate(II) with uranyl nitrate solution and in the reverse system, compounds of the composition $K_2UO_2[Fe(CN)_6]\cdot x[UO_2(NO_3)_2]$ were indicated, but in concentrated solutions the unassociated compound $K_2UO_2[Fe(CN)_6]$ was formed (Gaur *et al.*, 1953).

3.47.4 *Formation of uranium peroxy salts*

Rivenq (1945b), investigating the formation of the peroxy salts of uranium, titrated 100 ml containing 23·7 mmole of $UO_2(NO_3)_2$ with 2·52 N H_2O_2, using the early titration method. The titration graph was very curved. The start of the curved section was equivalent to the reaction $UO_3 + H_2O_2 \rightarrow UO_2(O_2) + H_2O$ and the end of the titration curve to the overall reaction $UO_3 + 2H_2O_2 \rightarrow UO_2(O_2)\cdot H_2O_2 + H_2O$ ($T = 0\cdot06$ degC/mmole).

3.47.5 *Determination of uranium(IV) as the sulphate complex*

The formation of sulphate complexes with uranium(IV) ions has been utilized by Guillot (1970) in an automated direct-injection method for uranium. A 20-ml aliquot of 3 M $(NH_4)_2SO_4$ was injected into a 2-ml sample of a solution containing from 10 to 300 g/l of UO_2^{2+} in 1 M acid, and the results obtained were correct to well within 1·5%.

Note

For the determination of free acidity in thorium and uranyl salts see Section 3.48.1(*viii*).

3.48 Neutralization reactions

The determination of acids or bases as such, or when present as minor components in other materials, is one of the ever-present needs in analysis, so it is not surprising that from its earliest days thermometric analysis has been applied to this determination.

The neutralization of strong inorganic acids in aqueous solution with strong bases is exothermic and the heat of reaction has a value that is very close to -13.5 kcal/mole irrespective of the ions involved. This constant value is that of the heat of formation of water from its ions, which are associated with the acid and base. A similar neutralization of a weak acid, i.e., one that is partially ionized, involves the continuous ionization of the acid during the neutralization so that the heat evolved from the formation of water is diminished by the heat required for the ionization of the acid (e.g., $+2.9$ kcal/mole for boric acid, determined thermometrically by Jordan and Dumbaugh, 1969b). It follows that the heat of neutralization of weak acids with strong bases is generally not reduced to less than -10 kcal/mole (e.g., -10.6 kcal/mole for boric acid). Similar arguments apply to the neutralization of weak bases with strong acids. The heat available for a thermometric neutralization titration is thus large enough for precise measurements to be made even of weak acids or bases at low concentrations.

The other factor affecting thermometric neutralization titrations is the pK of the acid or base (see Table 3.68). Jordan and Dumbaugh have stated that it is possible in the titration of an acid to determine the end-point with a precision of better than $\pm 1\%$ provided that:

  (*i*) the pK value is $\leqslant 10$
  (*ii*) the solution concentration is >0.003 M
  (*iii*) 100% excess of reagent is added to provide a sufficiently long titration curve after the end-point, for back-extrapolation.

Jordan and Henry (1966) investigating microtitration and micro-injection techniques stated that, in the mM concentration range for a strong acid or base neutralization reaction, it should be possible to obtain a precision and error of about $\pm 0.3\%$ and that the lowest concentration determinable within 1% is of the order of 0.001 M.

Table 3.68 gives the pK values of a number of mono- and polyprotic inorganic acids, and it can be seen that the majority of these can be titrated quantitatively and even those acids with ionization stages with pK values higher than 10 have given thermometric titration curves showing inflections for these ionizations. Again, similar arguments apply to weak inorganic bases, for example, ammonia (pK, 4.7) and hydrazine (pK, 5.7); to weakly acidic salts, (e.g., ammonium chloride, pK, 9.26) and to weakly basic salts.

If the pK values of two ionization stages have a difference of greater than about 2 then the thermometric titration curve should show two inflections caused by the separate neutralization of these stages. Under these circumstances it is therefore possible to distinguish in a single

Table 3.68 Values of pK for mono- and polyprotic inorganic acids at 18–25 °C

Substance	pK	Substance	pK
Nitrous acid	3·30	Pyrophosphoric acid (2)	2·28
Hydrazoic acid	4·82	(3)	6·70
Hypochlorous acid	7·40	(4)	9·37
Hydrocyanic acid	9·0	Chromic acid (2)	6·4
Boric acid	9·24	Carbonic acid (1)	6·5
Arsenious acid	9·26	(2)	10·2
Hydrogen peroxide	12	Sulphuric acid (2)	1·91
Hypophosphorous acid (2)	1·10	Sulphurous acid (1)	1·76
Phosphorous acid (1)	2·00	(2)	7·07
(2)	6·58	Hydrogen sulphide (1)	7·0
Orthophosphoric acid (1)	2·10	(2)	14·9
(2)	7·22	Arsenic acid (1)	2·25
(3)	12·36	(2)	6·77
Metaphosphoric acid	2·05	(3)	11·53

Note. The figures in brackets indicate the ionization step if the acid is polyprotic.

titration the separate ionization stages of a polyprotic acid or to distinguish between mixtures of acids.

The effect of the pK value is exemplified in Fig. 3.6 which shows in (a) the neutralization of diprotic sulphuric acid in which the two pK values

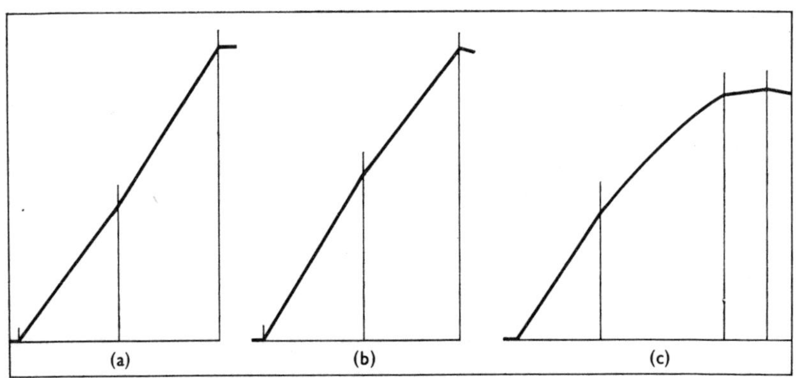

FIG. 3.6 Thermometric curves of the neutralization of inorganic mono- and polyprotic acids and their mixtures.
(a) sulphuric acid; Dutoit and Grobet (1921) early titration method;
(b) sulphuric and boric acids; Miller and Thomason (1959c);
(c) sulphurous and carbonic acids; Vaughan (1970).

differ by approximately 2 and it can be seen that a slight but sharp inflection occurs, showing the formation of the HSO_4^- ion; in (b) the neutralization of sulphuric acid and boric acid and of a mixture of the two, where it can be seen in the titration of the mixture that the inflection indicating the completion of the titration of the sulphuric acid and the beginning of that of the boric acid is sharper, indicating the larger difference in their pK values; in (c) the neutralization of sulphurous acid and carbonic acid, where both titration curves of the individual acids show inflections representing the first neutralization stages but in the titration of the mixed acids it can be seen that the second neutralization stage of sulphurous acid (pK 7·1) is titrated with the first neutralization stage of carbonic acid (pK 6·5), that is, the pK values are too close for the stages to be distinguished.

The discussion above implies that strong acids or bases are necessary as titrants for neutralization reactions and it is advisable to use them though results *can* be obtained by using weak acids or bases as titrants. For example: boric acid when titrated with ammonia solution, using the early titration method, gives a titration curve with an inflection indicating the formation of ammonium borate (Mondain-Monval and Pâris, 1938a); Ben-Yair (1957) suggested that bases could be titrated with solutions of zinc or cadmium nitrate, i.e., these salts act as weak acids. The scope of the thermometric method is thus wider than that of the neutralization methods in which visual indicators or potentiometric or conductimetric end-point detection are used and from which analytically acceptable values can only be obtained if one of the components of the system is a strong acid or base.

Table 3.69 Suggested titrants of the type HA $=$ A$^-$ $+$ H$^+$
(From Christensen *et al.* (1968c))

Titrant	Formula	Effective pK range
Hydrogen ion	H_3O^+	$<2·5$
Hydrazinium ion	$N_2H_5^+$	
Phosphoric acid	H_3PO_4	1–4
Anilinium ion	$C_6H_5NH_3^+$	3–6
Acetic acid	CH_3COOH	4–7
Pyridinium ion	$C_5H_6N^+$	
Dihydrogen phosphate ion	$H_2PO_4^-$	6–9
Imidazolinium ion	$C_2H_4NH^+$	
Acetylacetone	$CH_3COCH_2COCH_3$	8–11
Tris(hydroxymethyl)aminomethane	$(CH_3OH)_3CNH_3^+$	
Glycinium ion	$^+NH_3CH_2COO^-$	9–12
Monohydrogen phosphate ion	HPO_4^{2-}	11–14
Hydroxide ion	OH^-	

Christensen and his co-workers in a theoretical study (1968c) into the use of the thermometric method for the determination of the pK value for any proton dissociation reaction in the range <15 to >-1 suggested a number of titrants and their effective pK range of application, and these are given in Table 3.69.

In the discussion it is implied that the neutralization system should be liquid but this need not be the case, for it is possible to use a totally gaseous system as in the determination of ammonia with hydrogen chloride or a solid–liquid system as in the determination of the basicity of slag where the solid is added to an acid solution. These two methods are described below. Gas–liquid systems have also been described for the determination of (i) carbon dioxide, sulphur dioxide and nitrogen oxides by injection of gases containing them into alkali solutions (Zambonin and Jordan, 1969) and (ii) ammonia by continuous absorption in circulating acid (Guérin 1953). Further details of these determinations are given under the appropriate element.

3.48.1 *Titration of acids*

(i) SULPHURIC ACID

As discussed above, the first quantitative thermometric method described by Dutoit and Grobet (1921) showed that 10 ml of 0·4 N H_2SO_4 could be titrated with 7·5 N NaOH. The titration curve (see Fig. 3.6) showed a slight but sharp inflection caused by the neutralization of the first ionizable hydrogen and also that this neutralization was less exothermic than the second. Forman and Hume (1964) used acetonitrile as solvent and titrated sulphuric acid with 1,3-diphenylguanidine and obtained a precision of $\pm0·33\%$ and an error of $-0·5\%$. The titration curve showed a sharper inflection at the first neutralization stage because, in this case, the second neutralization was less exothermic than the first. Takeno (1964, 1965) determined the total acid content of fuming sulphuric acid by thermometric titration after first diluting the sample, which was weighed by means of a weight-pipette.

(ii) AMMONIUM HYDROGEN SULPHATE

Ammonium hydrogen sulphate in industrial concentrated sulphite and hydrosulphite solutions containing ammonium sulphate has been titrated continuously with a titrant prepared by diluting commercial ammonia with its own volume of water. Two ranges of concentration of up to 0·6 M and 7·5 M HSO_4^- were investigated and both gave linear calibration curves with a reproducibility of $\pm0·45\%$ (Štráfelda and Hájková, 1968).

(iii) HALOGEN ACIDS

The precision of the titration of hydrochloric acid in aqueous solution, using different types of apparatus, is summarized in Table 3.70. No precision values were quoted by Wasilewski and co-workers (1964) in their paper describing the development of the direct-injection method but they showed that between 41 and 207 μmole of HCl could be determined in 25 ml of solution by injecting 300 μl of 1 M NaOH.

Table 3.70 Precision of methods for the determination of hydrochloric acid

Hydrochloric acid Taken	Titrant Volume, ml	NaOH, M	No. of tests	Precision ±%	Apparatus	Ref.
50 mM	200	1·0	—	0·06	Burette; Beckmann thermometer	(a)
9–54 mmole	60	0·98	16	0·38	Constant-flow burette; thermistor	(b)
10 mM	—	1·0	—	2·5	Syringe; differential thermistor	(c)
2–7 mM	20	2·0	15	0·06	Direct injection; thermistor	(d)
2–40 μmole	2	1·0	6	1·0	Direct injection; thermistor	(e)
33 mM	30	1·0*	6	1·4	Automatic digital titrator	(f)
10–150 mM	—	—	28	4·4–1·5	Continuous flow titrator	(g)
5–50 mmole	—	0·6	6	2	Continuous flow titrator	(h)
0·1–1·6 mM	50	—	13	<0·8	Corrected non-adiabatic	(i)

* Sodium carbonate titrant
(a) Barthel *et al.* (1961); (b) Linde *et al.* (1953); (c) Takeuchi and Yamazaki (1969a);
(d) Avedikian (1966); (e) Jordan *et al.* (1966); (f) Priestley (1963a); (g) McLean and
Penketh (1968); (h) Priestley *et al.* (1965, 1968); (i) Nakanishi and Fujieda (1972).

Forman and Hume (1964) used acetonitrile as a solvent and titrated
hydrogen chloride, hydrogen bromide and perchloric acid in 15 ml of
solution with 1,3-diphenylguanidine and obtained the results given in
Table 3.71.

**Table 3.71 Titration of halogen acids in acetonitrile solution with
1,3-diphenylguanidine**
(Forman and Hume, 1964)

Acid	Amount present mmole	Precision ±%	Error %
Hydrogen chloride	0·603	0·47	−0·5
Hydrogen bromide	0·624	0·77	−0·9
Perchloric acid	0·348	0·13	−0·6

The early titration method (Barthel *et al.*, 1961), the direct-injection
method (Avedikian, 1966) and the 'corrected non-adiabatic' method
(Nakanishi and Fujieda, 1972) have all been used to determine the heats of
neutralization at infinite dilution of hydrochloric acid. The results for $\Delta H°$
obtained by the three methods were $-13·35 \pm 0·08$, $-13·55 \pm 0·05$ and
$-13·46 \pm 0·06$ kcal/mole respectively. These results compare extremely
well with those obtained by other methods; for example, those by

calorimetry range from $-13{\cdot}31$ to $-13{\cdot}50$ kcal/mole (Hale *et al.*, 1963; Vandersee and Swanson, 1963; Papee *et al.*, 1956), and those by electrometric methods range from $-13{\cdot}48$ to $-13{\cdot}52$ kcal/mole (Everett and Wynne-Jones, 1939; Harned and Robinson, 1940; Ackermann, 1958).

(*iv*) PHOSPHOROUS ACIDS

The quantitative titration of orthophosphoric acid was investigated by Dutoit and Grobet (1921), and the titration curve of $1{\cdot}7$ N H_3PO_4 with $5{\cdot}9$ M NaOH showed distinct inflections for all three neutralization stages of the acid, each having diminishing heats of neutralization; a similar titration by Barthel and co-workers (1961) gave the same shape of titration curve. Using continuous burette-addition and temperature-sensing with a Beckmann thermometer, it was shown that the precisions of the three end-points were $\pm 2{\cdot}8\%$, $\pm 2{\cdot}1\%$ and $\pm 0{\cdot}85\%$ respectively, but though the error of the first two end-points was only $-0{\cdot}3\%$, that for the third end-point was $+9\%$. Leydet and Champetier (1966), testing an electrovibratory stirrer, titrated 8 ml of $0{\cdot}1$ N H_3PO_4 continuously with 1 M NaOH at the rate of $0{\cdot}2$ ml/hour, and the titration graph of cal/hour *vs.* time showed three plateaux indicating the three neutralization stages of the acid, the third again being the least distinct. The values of log K and ΔH° for the neutralization of the HPO_4^{2-} ion have been determined by a single thermometric titration and were found to be $1{\cdot}61 \pm 0{\cdot}13$ and $-9{\cdot}14 \pm 0{\cdot}31$ kcal/mole respectively (Christensen *et al.*, 1966a).

If the pK values of the phosphorous acids given in Table 3.68 are examined it can be seen that not only is there a difference in the number of ionizable protons but that the pK values of the ionization stages differ considerably. This implies that individual acids in certain mixtures of the phosphorous acids may be determined by thermometric titrimetry. For example, Pâris and Robert (1943, 1946) titrated mixtures of meta-, pyro- and orthophosphoric acids with sodium hydroxide, using the early titration method. The first inflection in their titration curve was caused by the neutralization of the low pK (up to $2{\cdot}8$) ionization stages of the acids, i.e., all the meta- acid, half of the pyro-acid and one-third of the ortho-acid, the second inflection by the neutralization of the higher pK (up to $9{\cdot}4$) ionization stages, i.e., the remaining half of the pyro-acid, and the second third of the ortho-acid, and the final inflection, the least discernible, was caused by the neutralization of the remaining third of the ortho-acid. From these data the content of all three acids was calculated with an error of between 1 and 2%. Although, unlike volumetric indicator methods, it is not necessary to add barium chloride in order to obtain the third end-point for the orthophosphoric acid, it is better to do so. Under these conditions the results are more or less similar to those obtained by potentiometric or conductometric titrations.

Another example of this technique by Pâris and Tardy (1945, 1946) is the analysis of individual components in a mixture of hypophosphorous, phosphorous and orthophosphoric acids. Again three inflections were shown in the titration curve, the first representing all the hypophosphorous acid, half the phosphorous acid and one-third of the orthophosphoric acid, the second inflection the other half of the phosphorous acid and the second third of the orthophosphoric acid and the third inflection, again the least discernible, the final third of the orthophosphoric acid. The content of all three acids was calculated from the results, with an error of 1%. The results in this case show that the thermometric method is superior to others as it is more accurate and precise as well as being less time-consuming. It is not necessary to add barium chloride to obtain the final end-point for the ortho acid though it is better to do so.

Jordan and Dumbaugh (1959b) have determined the heats of neutralization and heats and entropies of ionization of hypophosphorous, phosphorous and orthophosphoric acids, and also of dihydrogen potassium phosphate.

(v) NITRIC ACID

The titration of nitric acid with 0·6 M NaOH has been carried out by using a continuous-flow titrator (Priestley *et al.*, 1965). Nitric acid in concentrations between 0·050 and 0·500 arbitrary units were determined within 1%.

Guillot (1970), investigating the performance of an automatic direct-injection apparatus, showed that nitric acid solutions in the range 0·3–3 M could be determined to well within 1·3% error when titrated with 0·2 M NaOH.

(vi) ARSENIC ACIDS

The monoprotic arsenic(III) acid was titrated by Mondain-Monval and Pâris (1938a) with 1·73 M NaOH, using the early titration method. The same authors (1938b) titrated the triprotic arsenic(V) acid, and the titration curve showed inflections for all three ionization stages of the acid, including the third with its high pK value of 11·53.

(vii) SELENIOUS ACID

It has been shown by Dastoor and Haldar (1967) that when selenious acid is titrated with approximately 1 N alkali two inflections appear in the titration graph, the second being sufficiently sharp for 20 mg of Se in 30 ml of solution to be determined with an error of $\pm$0·7%. In the presence of mineral acid, selenic acid or telluric acid only one inflection is obtained and this represents the total acidity.

(viii) BORIC ACID

As was shown in the earlier part of this section, the pK value and heat of neutralization of boric acid (9·24 and −10·6 kcal/mole) are such that it should be possible to determine boric acid by a thermometric titration. Mondain-Monval and Pâris (1938a), using the early titration method, carried out this determination with 1·73 M NaOH as titrant and although few details were given they also stated that it was possible to show the formation of ammonium borate. This was in direct contrast to the volumetric indicator method which cannot be used to determine boric acid with sodium hydroxide (and obviously not with ammonia) as titrant unless mannitol or another suitable polyhydroxy compound is added so that the complex so produced has a much lower pK value (approximately 4·2 as shown by Belcher *et al.*, 1970). The addition of mannitol (Millar and Thomason, 1959a) and of glycerol (Goyan *et al.*, 1961) in the thermometric titration of boric acid, as would be expected, increased the sharpness of the inflection at the end-point.

The precision of the titration of boric acid in aqueous solution when different types of apparatus are used is summarized in Table 3.72. No precision values were quoted by Wasilewski and co-workers (1964) in the paper describing their direct-injection method but they stated that between 1·95 and 9·92 mM boric acid could be determined in 25 ml of solution by the injection of 300 μl of 1 M NaOH.

Table 3.72 Precision of methods for the determination of boric acid

Boric acid Taken	Volume, ml	Titrant NaOH, M	No. of tests	Precision ±%	Apparatus	Ref.
1–100 mM	100	0·02–2	—	0·5	Burette; thermistor	(a)
0·1 mM	100	0·02	—	1·0	Burette; thermistor	(a)
10 mmole	60	0·98	4	0·38	Constant flow burette; thermistor	(b)
>3 mM	50	1·0	—	<1·0	Syringe; thermistor	(c, d)
0·102 mmole	5	1·0	5	1·69	Syringe; thermistor	(e)
0·3 mmole	50	2·0	3	1·1	Syringe; thermistor	(f)
0·05–0·50*	—	0·6	6	2	Continuous flow titrator	(g)

* Arbitary concentration units
(a) Pechar (1965); (b) Linde *et al.* (1953); (c, d) Jordan and Dumbaugh (1957, 1959a); (e) Miller and Thomason (1959a); (f) Harries (1968); (g) Priestley *et al.* (1965)

Pechar compared his thermometric titration with a potentiometric titration after the addition of mannitol; the former method gave results that proved to be both more accurate and more precise.

Harries (1968) not only titrated boric acid in aqueous medium but also in non-aqueous solution. In the non-aqueous titration two systems were studied: (*i*) methanol as solvent with sodium methoxide as titrant; (*ii*) benzene–methanol as solvent with tetra-n-butylammonium hydroxide as titrant. The accuracy, precision and heat of neutralization obtained in the titrations using these solvents as well as water are given in Table 3.73 in

Table 3.73 Precision, accuracy and heats of neutralization of boric acid
(Results using aqueous and non-aqueous media; from Harries (1968))

Solvent	Titrant	Ratio found/added	Precision ±%	$-\Delta H°$ kcal/mole
Water	2·0 M NaOH	1·014	1·1	10·2
Methanol	1·5 M NaOCH$_3$	1·004	0·23	13·9
Benzene–methanol	0·5 M (C$_4$H$_{11}$)$_4$NOH	0·995	0·65	13·6

which it can be seen that because of the higher heats of neutralization in the non-aqueous system, caused by the lower specific heats, a better accuracy and precision are obtained.

Miller and Thomason (1959c) titrated various mixtures of boric acid and sulphuric acid, keeping the concentration of one or the other component constant, and found that the results for either component were correct to within $\pm 3\%$. Similar results were stated to have been obtained for a boric–hydrochloric acid mixture.

Boric acid has also been determined as boron trifluoride (see Section 3.9.1(*i*)).

(*ix*) FREE ACID

Miller and Thomason (1959a) showed that it was possible to determine acids in the presence of hydrolysable cations. This 'free acid' was determined by titrating between 1 and 10 ml of the sample, diluted to 10 ml, with 1 M NaOH. Previous to this work no direct titration method existed for this determination although free acid could be estimated indirectly after the addition of complexing agents.

One of the solutions studied was that of zirconyl perchlorate containing hydrogen fluoride. It was found that a constant quantity of the hydrogen fluoride could be determined in the presence of a wide range of zirconyl concentrations, but uranium (IV), iron(II) and aluminium(III) were found to be additive interferents under these conditions, no inflections occurring in the titration curves between the free acid and the hydrolysis of these ions.

Other systems studied were: (*i*) the free nitric acid in thorium nitrate solution, either in the presence or absence of uranium(IV) ions, (*ii*) the free sulphuric acid in chromium(III) solutions and in uranium(VI) sulphate solutions. Addition of copper(II) ions to the latter solution produced a doubly inflected curve in which the copper(II) ions were shown to be hydrolysed after the titration of the free acid but before that of the uranium(VI). The results of Miller and Thomason's work showed that milliequivalents of free acid could be determined with a relative standard error of ± 1–2% at the 0·1 meq level and $\pm 5\%$ at the 0·01 meq level.

Vaughan (1966) showed that free sulphuric acid in sulphuric extracts of tar oils, i.e., extracts containing pyridine-base sulphates, could be titrated with 5 M NaOH (for results see Table 4.11).

(*x*) WEAK-BASE SALTS

As in indirect volumetric titrations, it is possible to titrate weak-base salts with a strong base, using the thermometric technique. This 'displacement' reaction is exemplified in the titration of ammonium chloride (having a pK value of 9·26) with sodium hydroxide, as described by Linde and his co-workers (1953). Using a continuous-flow burette and thermistor apparatus, they obtained a small temperature rise ($\Delta H \approx -0\cdot25$ kcal/mole) and the inflection at the end-point was correspondingly slight but sharp. As would be expected from the pK value they also showed that hydrochloric acid, if present, was titrated first, followed by the ammonium chloride.

They obtained a precision of $\pm 0.2\%$ when hydrochloric acid or ammonium chloride was titrated with 0·95 M NaOH but for mixtures the precision was poorer, $\pm 1.0\%$.

Other 'displacement' titrations, in which strong-base titrants are used, can be found under the appropriate cation heading and include the titration of the nitrate and other salts of aluminium, copper, lead, magnesium and zinc. These titrations, though used in the study of the intermediate compounds formed in the titration, can also be used to determine the particular metal or its salt.

3.48.2 *Titration of bases*

(*i*) SODIUM AND POTASSIUM HYDROXIDE

A number of different methods have been used to determine sodium hydroxide in aqueous solution and the precision of these methods is given in Table 3.74. Snyder (1968) determined up to 4% sodium hydroxide in

Table 3.74 Precision of methods for the determination of sodium hydroxide and ammonia by titration with acid

Substance Taken	Volume, ml	Titrant HCl, M	No. of tests	Precision $\pm\%$	Apparatus	Ref.
SODIUM HYDROXIDE						
4·902 mmole	55	1·032	22	0·17	Constant-flow burette, thermistor	(*a*)
4·902 mmole	155	1·032	5	0·60		
1·250 mmole	—	0·25	16	0·4	Derivative titrator	(*b*)
2·500 mmole	—	0·50	6	0·2		
0·46–3·3 mmole	75	1·013	—	0·1	Differential titrator	(*c*)
M/60	30	1·0	10	0·6	Automatic digital titrator	(*d*)
0·1 M	20	1·0	11	0·4	Automatic titrator	(*e*)
0·025–0·50*	—	0·6	6	1·0†	Continuous-flow titrator	(*f*)
2·7–14·9 g/l	—	1·0	—	0·9	Continuous-flow titrator	(*g*)
AMMONIA						
8·54 mmole	60	1·008	5	0·56	Constant-flow burette, thermistor	(*a*)
0·3932 mmole	75	0·97	5	0·8	Differential titrator	(*c*)
M/60	30	1·0	10	1·1	Automatic digital titrator	(*d*)

*Arbitrary concentration units. † Similar precision obtained in the titration of potassium hydroxide
(*a*) Linde *et al.* (1953); (*b*) Zenchelsky and Segatto (1957b); (*c*) Tyson *et al.* (1961); (*d*) Priestley (1963a); (*e*) Priestley *et al.* (1963); (*f*) Priestley *et al.* (1965, 1968); (*g*) Štráfelda and Kroftová (1968a)

non-aqueous process streams, using a continuous-flow titrator. He employed 1% acetic acid in propan-2-ol as titrant and obtained results correct to within $\pm 0.05\%$ NaOH, which he expected to improve with continuous monitoring on streams of better uniformity.

Jahr and co-workers (1968) titrated 1·5 ml of 0·01 M KOH in 80 ml of solution with 0·05 N H_2SO_4 and obtained a relative standard deviation of ±0·2%.

(*ii*) AMMONIA

Ammonia has also been determined in aqueous solution by titration with acid and the precision obtained by different methods is given in Table 3.74. Hume and Duffield (1964) determined 0·1–1·0% by volume of ammonia in air by titration with gaseous hydrogen chloride. Ammonia has been titrated with 0·1 M HBr in the non-aqueous solvent acetonitrile (Forman and Hume, 1964). In four replicate titrations of 1·320 mmole of NH_3 in 15 ml of solvent, an accuracy of 99·7% and a precision of ±0·95% were obtained. This precision was not as good as that for a comparable aqueous titration and this was attributed to volatilization from the solvent.

(*iii*) WEAK-ACID SALTS

The thermometric titration of weak-acid salts with strong acids is analogous to the titration of weak-base salts with strong bases above. A number of these 'displacement' reactions have been studied in aqueous solution and the details and results are given in Table 3.75.

Table 3.75 Details of methods for the determination of various weak-acid salts by titration with strong acid

Substance Taken	Volume, ml	Titrant HCl, M	No. of tests	Precision ±%	Apparatus	Ref.
SODIUM CARBONATE						
5·0 mmole	60	1·032	6	0·31	Continuous-flow burette; thermistor	(*a*)
M/60	30	1·0	10	0·5	Automatic digital titrator	(*b*)
1·0 mmole	20	1·0	—	—	Rapid automatic titrator	(*c*)
SODIUM BICARBONATE						
M/60	30	100	7	0·3	Automatic digital titrator	(*b*)
SODIUM TETRABORATE						
M/60	30	1·0	6	0·8	Automatic digital titrator	(*b*)
1·0 mmole	20	1·0	—	—*	Rapid automatic titrator	(*c*)
TRISODIUM ORTHOPHOSPHATE						
1·4 M	15	6·6†	—	—§	Early titration method	(*d*)
M/60	30	1·0	10	0·6	Automatic digital titrator	(*b*)
0·025–0·20‡	—	0·6	5	2·0	Continuous-flow titrator	(*e*)
TETRASODIUM EDTA						
M/120	30	1·0	7	0·2	Automatic digital titrator	(*b*)

* Poor end-point † Nitric acid § Four distinct end-points
‡ Arbitrary concentration units
(*a*) Linde *et al.* (1953); (*b*) Priestley (1963a): (*c*) Priestley *et al.* (1963); (*d*) Dutoit and Grobet (1921); (*e*) Priestley *et al.* (1965, 1968)

A continuous-flow titrator has been used in the titration of sodium carbonate and sulphite with 0·6 M HCl (Reference (*e*) Table 3.75) and it is obvious that there are a great number of weak-acid salts not given in Table 3.75 that can be titrated. A wider range of displacement reactions is possible in non-aqueous solution but there is a limitation because of the low solubility of inorganic salts. For example, Keily and Hume (1964), using anhydrous acetic acid as the medium, titrated a number of salts at 0·05 M concentration with 0·5 M $HClO_4$. Sharp end-points were obtained in the titration of tri-, di- and monosodium orthophosphates and of the nitrates of lithium and sodium.

In the titration of trisodium phosphate, using the early titration method, four end-points were observed, the first of these being caused by the neutralization of the 'free' base present in the salt. This shows that free base or added excess of base can be determined in the presence of and

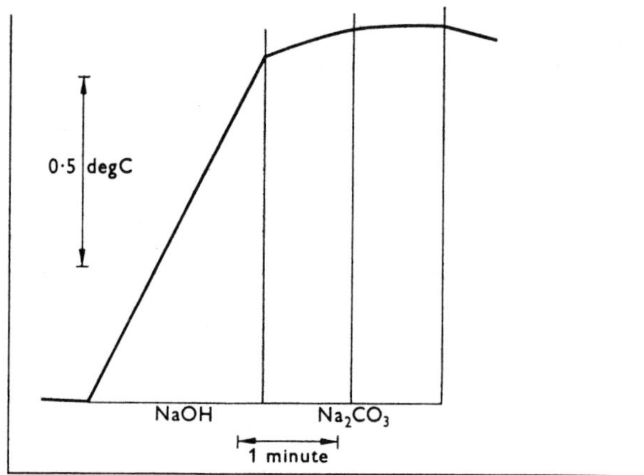

FIG. 3.7 Titration of a mixture of sodium hydroxide and sodium carbonate.
5 meq. NaOH + 5 meq. Na_2CO_3 in 50 ml; from Linde *et al.* (1953).

together with the weak-acid salt. This is exemplified in the work of Linde and co-workers (1953) who titrated sodium hydroxide and sodium carbonate in admixture. Figure 3.7 shows the titration curve obtained and it can be seen that not only is the sodium hydroxide (i.e., the 'free' base) titrated with a sharp end-point, but both end-points of the titration of the sodium carbonate are discernible.

Another example of a more complex neutralization that can be carried out by thermometric titration is shown in Fig. 3.8. In this a mixture of equimolar amounts of sodium hydroxide, sodium carbonate and sodium sulphite is titrated with 5 M HCl. It can be seen that the titration curve has five inflections showing in order the neutralization of (*i*) the sodium hydroxide (the 'free' base); (*ii*) the first and second ionizations of the sodium carbonate; (*iii*) the first and second ionizations of sodium sulphite (Vaughan, 1970).

The determination of the basicity of slags is another example of the determination of free basicity in the presence of weak-acid salts, in this case the calcium salts of phosphoric and silicic acids. The slag was ground to pass a 0·1-mm mesh sieve and 1 g was then added carefully to 50 ml of hydrochloric acid in a Dewar flask so that the slag particles did not cohere because of the formation of silicic acid. The basicity calculated from the temperature rise of the acid (about 9 degC) was within ±1%

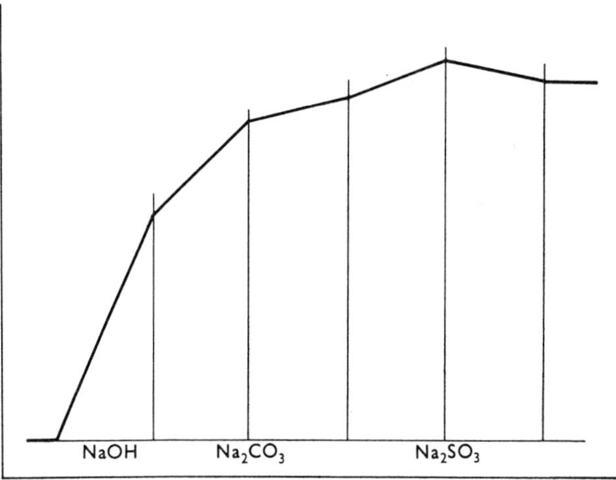

FIG. 3.8 Titration of a mixture of sodium hydroxide, sodium carbonate and sodium sulphite.
0·5 mmole of each in 10 ml; from Vaughan (1970).

(Mandl *et al.*, 1962; Sajó, 1957). The solution remaining after measurement was used for the determination of the silica and iron contents of the slag (see Sections 3.15.1 and 3.36.1(*iv*) respectively).

The values of log K and $\Delta H°$ for the reaction $HSO_4^- \rightleftharpoons H^+ + SO_4^{2-}$ have been determined in a single titration and were found to be, respectively, $1·91 \pm 0·01$ and $-5·64 \pm 0·08$ kcal/mole (Christensen *et al.*, 1966a) and also $1·930 \pm 0·005$ and $-5·41 \pm 0·04$ kcal/mole (Wanders and Zweitering, 1969).

3.49 Metal-ion complexes of aminopolycarboxylic acids

There is considerable interest in the formation of metal complexes involving aminopolycarboxylic acids. A number of these have been studied by thermometric titrimetry and include ethylenediaminetetra-acetic acid, EDTA; di(2-aminoethoxy)ethanetetra-acetic acid, EGTA; ethylenediaminediacetic acid, EDDA; and nitrilotriacetic acid, NTA.

3.49.1 *Metal–EDTA complexes*

Ethylenediaminetetra-acetic acid (EDTA) has been widely studied as a thermometric titrant for metals, paralleling its extensive use in other types of volumetric analysis. Although the specific use of EDTA as the disodium or tetrasodium salt for the thermometric determination of individual metals is detailed under the appropriate metal (see Table 3.76) the methods given do not by any means cover the wide range of metals that may be determined with it. Table 3.76 gives the heats of formation of EDTA chelates of the metals that have been determined thermometrically and also those determined by other means, together with the appropriate stability constants ($\log K$). The table should help those who wish to develop thermometric methods using EDTA, as it shows which metals are likely to interfere and it is for this reason that the tabulation is given in decreasing values of the stability constant.

In general, the method is applicable to the determination of concentrations as low as 0·5 mM with an error within 3%; a better accuracy (error

Table 3.76 Thermodynamic data for metal–EDTA chelate formation
$$M^{x+} + EDTA^{4-} \rightleftharpoons [M.EDTA]^{(4-x)-}$$

Cation	Stability constant $\log K$	$\Delta H°$ values, kcal/mole at 20–25 °C					
		(i)	(ii)	(iii)	(iv)	(v)	(vi)
Fe^{3+}	25·1	—	—	—	—	—	—
In^{3+}	25·0	—	—	—	—	7·23	—
*Hg^{2+}	21·8	—	—	—	—	—	−18·9
*Sn^{2+}	—	−nq	—	—	—	—	—
*Cu^{2+}	18·8	−8·3	−8·2	−8·7	−8·2	−8·67	−8·2
*Ni^{2+}	18·5	−7·2	−7·6	−8·4	−7·4	−8·35	−7·6
Y^{3+}	18·1	—	—	—	—	−0·32	—
*Pb^{2+}	18·0	−12·7	−13·1	−14·1	−12·8	−13·20†	—
Gd^{3+}	17·0	—	—	—	—	−1·11	—
*Cd^{2+}	16·5	−9·7	−9·1	−10·1	−9·2	−10·08	−9·1
*Zn^{2+}	16·5	−4·5	−4·5	−5·6	−4·6	−5·61	−4·9
Nd^{3+}	16·5	—	—	—	—	−2·98	—
*Al^{3+}	16·1	+10·9	—	—	—	+12·58	—
*Co^{2+}	16·1	−3·5	−4·0	—	−4·2	−4·40	−4·2
Ce^{3+}	16·0	—	—	—	—	−2·43	—
Fe^{2+}	14·4	—	—	—	—	—	−4·0
UO_2^{2+}	—	+nq	—	—	—	—	—
*Mn^{2+}	13·8	−6·4	−5·2	—	—	−5·45	−4·6
*Ca^{2+}	10·7	−5·6	−5·8	−6·5	−5·7	−6·45	—
*Be^{2+}	—	+2·3	—	—	—	—	—
*Mg^{2+}	8·7	+4·8	+3·1	+3·1	+5·5	+3·14	—
Cr^{3+}	—	+7·3	—	—	—	—	—
Sr^{2+}	8·6	−4·9	−4·2	−4·1	—	−4·11	—
*Ba^{2+}	7·8	−4·6	−5·1	−4·8	—	−4·83	—
*Ag^+	7·3	—	—	—	—	—	—
NH_4^+	—	+4·8	—	—	—	—	—
*Li^+	2·8	+2·2	+0·1	—	—	—	—
K^+	—	+nq	—	—	—	—	—
La^{3+}	—	−3·9	—	—	—	—	—
Na^+	—	—	−1·4	—	—	—	—

* Further details are given in the text under the individual metal
† This value also quoted by Jordan *et al.* (1966): micro direct injection
(*i*) Priestley *et al.* (1956); continuous-flow enthalpimetry
(*ii*) Charles (1954); direct injection
(*iii*) Care and Staveley, (1956); internal injection
(*iv*) Jordan and Alleman (1957); thermometric titration
(*v*) Sillén and Martell (1967); compilation of data
(*vi*) Brunetti *et al.* (1969); thermometric titration
nq Priestley *et al.* (1963); automatic titration, absolute value not quoted, only sign

within 1%) can be obtained with a 10 mM solution (Alleman, 1957), though this is only achieved by the use of reasonably sensitive apparatus because of the relatively low heats of formation (only that of lead is equivalent to a strong acid/base neutralization). However, some metals, e.g. chromium, react only slowly with EDTA and, therefore, cannot be determined directly.

As well as determining the free energy and enthalpy changes accompanying the reaction of the bivalent metals with EDTA (given in column (vi) of Table 3.76), Brunetti and co-workers also determined these values for the protonation of the complexes, i.e., $MY^{2-} + H^+ \rightleftharpoons MHY^-$.

(i) MULTI-ION AND SUBSTITUTION TITRATIONS

The stability constants given in Table 3.76 show whether a particular mixture of ions can be titrated consecutively in a single titration. This is exemplified in the titration of a mixture of calcium and magnesium. The calcium has the higher constant and therefore will be titrated first; the constant is sufficiently higher than that for magnesium for there to be a distinct inflection in the titration curve, but because the magnesium–EDTA reaction is endothermic the end-point between the two ions is considerably sharpened, as can be seen in Fig. 3.1 (see also Section 3.5.1).

An example of a titration of a complex mixture is that where iron, aluminium, calcium, magnesium and barium ions were distinguished in a single titration using an automatic titrator (Priestley $et\ al.$, 1963). Though the end-points were reasonably sharp, they did not correspond exactly to the expected values for the ions, and this was attributed to interactions taking place under the experimental conditions used.

That this discrepancy was caused by substitution reactions was shown by the same workers in the titration of the EDTA metal complexes with other metal ions. If the calcium chelate were titrated with M Cu^{2+} an exothermic curve was obtained with a rather poor end-point, showing that the heat of chelation of EDTA with copper was greater than that with calcium. A similar titration of the magnesium chelate gave a sharp end-point and a much more exothermic reaction whilst titrations of the calcium and magnesium chelates with M Al^{3+} gave two endothermic reactions, the calcium in this case, as would be expected, giving the more endothermic.

From this work it was concluded that analytical methods could be devised for a metal having a low heat of chelation with EDTA if this metal could be used to displace another metal, suitably chosen so as to give a much greater change of temperature, from its EDTA chelate.

3.49.2 *Metal–EGTA complexes*

Boyd and his co-workers (1965) determined the heats of formation of the 1 : 1 complexes of EGTA with six bivalent metals; the results are shown in Table 3.77. The high heat of formation found with cadmium implies that it should be possible to develop a precise method for its determination.

3.49.3 *Metal–EDDA complexes*

The enthalpies of formation of the 1 : 1 EDDA complexes with bivalent metals have been determined by Degischer and Nancollas (1970). They

Table 3.77 Heats of formation of bivalent metal–EGTA complexes
(From Boyd *et al.* (1965))

Metal ion	ΔH, kcal/mole
Mg^{2+}	$+5{\cdot}49 \pm 0{\cdot}12$
Ca^{2+}	$-7{\cdot}94 \pm 0{\cdot}07$
Sr^{2+}	$-5{\cdot}74 \pm 0{\cdot}15$
Ba^{2+}	$-8{\cdot}99 \pm 0{\cdot}22$
Zn^{2+}	$-5{\cdot}02 \pm 0{\cdot}10$
Cd^{2+}	$-14{\cdot}89 \pm 0{\cdot}20$

also determined these values for the $1:1:1$ $M^{2+}:EDDA:$ ethylene-diamine complexes. The results obtained for the $1:1$ $M^{2+}:EDDA$ complexes are given in Table 3.78.

Table 3.78 Heats of formation of bivalent metal–EDDA complexes
(From Degischer and Nancollas (1970))

Metal ion	ΔH, kcal/mole
Mn^{2+}	$-0{\cdot}85 \pm 0{\cdot}1$
Co^{2+}	$-5{\cdot}76 \pm 0{\cdot}1$
Ni^{2+}	$-10{\cdot}0 \pm 0{\cdot}2$
Zn^{2+}	$-6{\cdot}1 \pm 0{\cdot}1$
Cd^{2+}	$-5{\cdot}38 \pm 0{\cdot}1$

3.49.4 *Metal–NTA complexes*

Thermometric titrimetry has been used by Freeberg (1969) to study the bivalent metal nitrilotriacetate chelation reaction. The metals studied were Cu(II), Zn(II), Ni(II) and Co(II), and 1 M solutions of these were used to titrate approximately 4 mmole of nitrilotriacetate (NTA).

The pH of the NTA solution affects the ions present; if the pH is between 1 and 5, $HNTA^{2-}$ and H_2NTA^- are present, and if it is between 8 and 12, $HNTA^{2-}$ and NTA^{3-} are present and this affects the thermometric titration results. It was found that in acid media at pH less than 5 the titration curves were endothermic and sharp but at pH values higher than 5, where the H_2NTA^- ion was predominant, the end-points were very curved, but as the solution became more basic the titration curve changed and showed an initial exothermic section followed by an endothermic section. The exothermic section was attributed to the metal(II)–NTA^{3-} reaction and the endothermic section to the sequential metal(II)–$HNTA^{2-}$ reaction. Further proof of this reaction was shown by titrating NTA at high pH with Zn(II), alone, and after the addition of two equivalents of Mg(II). When the magnesium ions were present the reaction was considerably more exothermic and showed only one end-point.

Note

The determination of EDTA and of traces of metal ions by using EDTA as an intermediate in a catalytic indicator method is given in Section 3.52.

3.50 Metal alkyls

Metal alkyls are becoming increasingly important to the chemical industry as catalysts of the Ziegler type, and their analysis by thermometric titrimetry has been shown to be both easier and quicker than by previous methods and equally satisfactory.

3.50.1 *Lithium alkyls*

The content of s-butyl-lithium in hydrocarbons has been determined by titrating the sample, diluted to 50 ml with toluene, with 1 M butanol in toluene (Everson, 1964). All the reagents had to be dry, and a dry nitrogen purge was maintained in the titration vessel during the titration. The reaction, a displacement one forming lithium butoxide, is highly exothermic, and this coupled with the low specific heat of the solvents gave a high T value of 2·8 degC/mmole. Contents as low as 0·01% of butyl-lithium can therefore be determined in a 50-ml sample. The results obtained are compared in Table 3.79 with those obtained by a vanadium pentoxide oxidation method.

Table 3.79 Determination of s-butyl-lithium in hydrocarbons
(Comparison of results; from Everson (1964))

Sample	*Vanadium oxidation method* %	*Thermometric method* %
A	12·37, 12·50	12·66, 12·70
A after standing	11·74, 11·88	11·88
A diluted	0·80	0·83, 0·84, 0·85

The method, unlike chemical methods, is not affected by the presence of alkoxides, because they are formed in the titration, and it should be applicable to other lithium alkyls.

3.50.2 *Aluminium alkyls*

Alkyl aluminiums, alkylaluminium hydrides and halides are Lewis acids and readily form electron-sharing complexes with Lewis bases such as amines, ethers, ketones and pyridines. These reactions are exothermic with heats of reaction in the range −10–25 kcal/mole. They also react with alcohols, which replace the alkyl groups successively, and in this case the reactions are very exothermic with heats of reaction in the range −40–50 kcal/mole. These reactions are therefore suitable for thermometric titrimetry and methods based on them have been developed by Hoffmann and Tornau (1962) and by Everson and Ramirez (1965a).

(*i*) DETERMINATION OF ALUMINIUM ALKYLS WITH AMINES

Hoffman and Tornau titrated the sample, dissolved in 50 ml of cyclohexane, with triethylamine (also in solution in cyclohexane) and showed that aluminium trialkyls were titrated first and distinguished from alkylaluminium hydrides and halides. The results are summarized in Table 3.80. Everson and Ramirez also investigated the use of *N*-methylaniline and benzalaniline as titrants. All three titrants formed 1 : 1 titrant : aluminium alkyl complexes, but the titration curve with benzalaniline showed considerable curvature that was attributed to the formation of a 1 : 2 titrant : trialkylaluminium complex.

Table 3.80 Titration of aluminium alkyls with triethylamine
(From Hoffman and Tornau (1962))

Substance	R_3Al, %	R_2AlH, %
Triethylaluminium	98·0	0·3
,,	89·2	8·1
Diethylaluminium chloride	98·1	nil
,, ,,	95·3	2·7
Dipropylaluminium hydride	7·0	93

(*ii*) DETERMINATION OF ALUMINIUM ALKYLS WITH ETHERS

It was found that ethers such as di-n-butyl ether did not react with alkylaluminium hydrides and consequently triethylaluminium could be determined in the presence of diethylaluminium hydride or of dipropylaluminium hydride by titration in cyclohexane or toluene solution.

(*iii*) DETERMINATION OF ALUMINIUM ALKYLS WITH KETONES

Both acetone and benzophenone have been investigated as titrants and were found to react only with the dialkylaluminium hydrides, but the stoichiometry depended on the presence of ether. In its presence a 2 : 1 ketone : aluminium hydride complex was formed but in its absence there was a 1 : 1 association.

(*iv*) DETERMINATION OF ALUMINIUM ALKYLS WITH PYRIDINES

With pyridine, isoquinoline and 2, 2′-bipyridyl as titrants it was found that trialkylaluminiums could be distinguished from dialkylaluminium hydrides in a single titration, the hydrides being titrated after the trialkyls. The association of isoquinoline with the hydride was in the ratio of either 1 : 1 or 2 : 1 and that of 2, 2′-bipyridyl with the trialkyls in the ratio 1 : 2 with the normal alkyls and 1 : 1 with the isoalkyl isomers, otherwise all ratios were 1 : 1.

(*v*) DETERMINATION OF ALUMINIUM ALKYLS WITH ALCOHOLS

The reaction of alcohols with trialkylaluminiums $Al(Alk)_3$ proceeds in three stages forming (*i*) $Al(Alk)_2OR$, (*ii*) $Al(Alk)(OR)_2$ and finally (*iii*)

Al(OR)$_3$, with high but diminishing heats of reaction at each stage. Only the first two stages are useful analytically, and, for example, in the titration of triethylaluminium in 52 g of cyclohexane with 5% 2-ethylhexanol also in cyclohexane, a Ziegler activity of 89·8% was obtained using the stage (*i*) end-point which was not well-defined, and 91·0% using the stage (*ii*) end-point which was sharp ($T = 2·4$ degC/mmole, for stages $i + ii$). Methanol or ethanol in solution in benzene or toluene were also used as titrants. It was found that with secondary alcohols the stage (*ii*) end-point was much less well-defined and with tertiary alcohols it was non-existent. The stage (*i*) end-point with tertiary alcohols was very sharp and consequently tertiary butyl alcohol has been extensively used as a titrant for both aluminium trialkyls and for dialkylaluminium hydrides and halides.

(*vi*) DETERMINATION OF ALUMINIUM ALKYLS WITH OXINE

The use of 8-quinolinol (oxine) as a titrant for aluminium alkyls and also zinc alkyls (Section 3.50.3) is interesting as an example of the wide range of materials that can be used as titrants in thermometric analysis.

The reaction in this case is a chelation and the compounds formed are more complex. Ratios of oxine to trialkylaluminiums and also to dialkylaluminium hydrides and halides of 1 : 1, 2 : 1 and 3 : 1 were found. Unlike other titrants, oxine was found to titrate dialkylaluminium alkoxides, forming both 1 : 1 and 1 : 2 chelates. Unfortunately, of the end-points in this titration only that representing the sum of Al(Alk)$_3$ and Al(Alk)$_2$H (or Al(Alk)$_2$Cl) contents was well-defined. In the titration of the alkoxides, the final reaction in which the alkoxide group was removed was found to be very slow. It was thought that this potentially valuable titration could be improved by a slower titration at higher temperatures.

(*vii*) ASSAY OF COMMERCIAL ALUMINIUM ALKYLS

From the work described in the sections above, Everson and Ramirez concluded that three titrants, comprising an ether, a ketone and an alcohol, would be adequate for the analysis of most commercial aluminium alkyls in the C$_1$–C$_8$ alkyl range.

Titrations with di-n-butyl ether and with benzophenone were used to obtain values for Al(Alk)$_3$ (or Al(Alk)$_2$Cl) and for Al(Alk)$_2$H. If their sum, i.e., their 'Ziegler' activity, were of prime interest then a titration with tertiary butyl alcohol was sufficient. All three titrants are available in pure form and are readily soluble in toluene to yield the 1–2 M concentration required for the titrants.

Water and air had to be excluded during the titration by nitrogen purging but Everson and Ramirez also noted that they had also to be eliminated from the solvents and to do this they used a small quantity of sample as a clean-up reagent before a succession of sample titrations.

3.50.3 *Zinc alkyls*

Bipyridyl (BIPY) and *o*-phenanthroline (PHEN) form 1 : 1 complexes with diethylzinc and have been used as titrants in its analysis (Everson and Ramirez, 1965b). Anisole was used as the solvent for the substances at between 0·8 and 1·6 M concentration. The reaction was exothermic with an estimated heat of reaction of -10 ± 2 kcal/mole for both titrants. Of the two titrants PHEN gave the sharper end-points and was the one preferred.

Oxine (8-quinolinol) has been investigated as a titrant for zinc alkyls as well as for aluminium alkyls (see above) and in this case was more successful. The exothermic heat of reaction was greater than with BIPY and PHEN, being about -33 ± 4 kcal/mole. The titration curve of diethyl zinc with oxine as titrant, both being in solution in toluene, showed three inflections. The first corresponded to a 1 : 1 reaction; the second, a less well defined inflection which was not always present, could not be correlated with a specific reaction; the third corresponded to a 2 : 1 oxine : zinc-alkyl reaction.

The results given in Table 3.81 show that both PHEN and oxine gave precise results which compared well with results obtained by hydrolysis of the zinc diethyl followed by an EDTA titration.

Table 3.81　Assay of zinc diethyl
(From Everson and Ramirez (1956b))

Titrant	Content mmole/g
PHEN	1·134
	1·137
Oxine	
1st end-point	1·137
	1·115
2nd end-point	1·142
	1·140
EDTA (indicator)	1·124
	1·131

Precautions similar to those described in Section 3.50.2(*vii*) above had to be taken.

3.51　Distribution systems

Information concerning the number of organic and inorganic ligands associated with an extracted compound has been obtained by thermometric titrimetry. Keltrup and Specker (1967) used cyclohexane, isobutyl methyl ketone and dibutyl phosphate as solvents in the thermometric study of two distribution systems. The extraction of iron from a solution containing potassium thiocyanate, and the distribution of bismuth, cadmium, mercury and zinc ions in a solution containing lithium iodide were investigated.

3.52　Catalytic indicator methods

The catalytic end-point indicator methods (see Section 1.4.2) developed by Professor Weisz and his co-workers for the determination of inorganic substances are not specific but can be used to determine very small quantities of anions or cations. So far, two highly exothermic catalytic systems have been utilized; the cerium(IV) oxidation of arsenic(III), catalysed by iodide ions (Weisz *et al.*, 1969; Burton and Irving, 1970; Kiss, 1970) and the decomposition of hydrogen peroxide or the hydrogen peroxide oxidation of resorcinol, catalysed by manganese(II) ions (Weisz and Kiss, 1970; Kiss, 1970).

3.52.1 *Iodide catalyst*

Only those iodides are suitable that form rapidly and completely and are sufficiently insoluble to ensure that the free iodide concentration derived from them in the solution is very small, thereby giving no catalytic effect. Silver(I), mercury(II) and palladium(II) iodides fulfil these requirements,

Table 3.82 Determination of Ag^+, Hg^{2+} and Pb^{2+} by the catalytic indicator method

(From Weisz *et al.* (1969))

Metal ion determined	General Method		Accurate Method		
	Added μg	*Found* μg	*Added* μg	*Found* μg	*Precision* $\pm \mu g$
Silver(I)	53·9	53·9	1·93	1·91	0·01
	107·9	108·5	4·10	4·06	0·02
	214·8	212·8	420·7	419·1	0·7
	429·5	424·7	431·5	431·1	1·2
	539·4	543·8			
Mercury(II)	100·3	100·9	1·34	1·34	0·01
	200·6	198·0	4·04	4·02	0·02
	401·2	405·7	180·5	181·3	0·2
	601·8	601·5	318·0	318·8	0·5
	802·4	811·5			
	1003	991·0			
Palladium(II)	53·2	53·2	1·32	1·31	0·01
	106·4	106·7	4·32	4·30	0·02
	212·8	215·7	160·3	159·6	0·7
	319·2	324·3	280·4	281·4	0·8
	425·6	433·3			
	532·0	538·9			

and therefore these metal ions can be determined directly or can be used as intermediate reagents in the determination of anions which react with them to form insoluble products, i.e., in a back-titration procedure.

In the general method, 0·5 ml of test solution is pipetted into a 10–ml Dewar flask followed by 1 ml each of 2 N H_2SO_4 and of water and lastly 0·1 ml each of 1·3 N As(III) at pH 9 and 0·4 M Ce(IV) in 1 N H_2SO_4. The solution is then titrated with 0·01 M KI at a rate of about $50 \mu l$/min. In the back-titration procedure a known volume of 0·01 M KI that is equivalent to 1·5–2 times the amount present of the anion to be determined is added after the test solution.

In a more accurate method, a first titration as above gives a rough value. To a second test solution 90% of the required volume of 0·01 M KI is added before the As(III) and Ce(IV) are added, and the titration is completed with 0·001 N KI as the titrant.

Table 3·82 shows the results obtained by the general and accurate methods for the direct determination of silver(I), mercury(II) and palladium(II); the precision figures quoted were the result of nine replicate determinations.

Table 3.83 shows the results obtained in the back-titration method, using the three cations as intermediates for the determination of Cl⁻, Br⁻, I⁻, CN⁻, SCN⁻, S²⁻, and [Fe(CN)₆]⁴⁻. Though silver and palladium form insoluble sulphides these are oxidized by the Ce(IV) ion and these metals cannot be used for this determination. When mercury acetate is

Table 3.83 Determination of anions by a back-titration catalytic indicator method
(From Weisz *et al.* (1969))

Anion	Taken	Intermediate added		
		Mercury(II) *acetate* Anion found	*Silver nitrate* Anion found	*Palladium*(II) *chloride* Anion found
	μg	μg	μg	μg
Cl⁻	17·73	17·70	17·61	—
	35·46	—	35·32	—
	70·91	69·93	71·80	—
	106·4	—	107·9	—
	117·3	118·7	—	—
	177·3	—	179·3	—
Br⁻	39·96	39·56	39·56	—
	159·8	161·8	161·9	—
	399·6	405·2	404·4	—
I⁻	63·45	63·39	63·29	63·45
	253·8	253·8	254·2	253·8
	634·5	636·1	633·9	634·5
CN⁻	13·01	13·09	—	12·94
	52·04	51·36	—	52·41
	130·1	131·9	—	129·1
SCN⁻	29·04	28·64	28·66	28·79
	116·2	117·8	115·2	117·7
	290·4	288·2	294·6	293·8
S²⁻	16·03	15·96	—	—
	64·13	64·35	—	—
	160·3	159·2	—	—
[Fe(CN)₆]⁴⁻	105·8	104·8	—	—
	423·9	429·0	—	—
	1059	1055	—	—

used for the determination of Cl^- and Br^- the 2 N sulphuric acid should be omitted from the titrand.

Burton and Irving (1970) used a similar catalytic system, but their technique was different for they allowed the titration to proceed until an appreciable temperature rise had occurred (about 30 s after the true end-point). The syringe motor was then switched off and the temperature rise allowed to continue under the catalytic effect of the iodine excess and was recorded to give a linear curve after the end-point. Extrapolation of the linear curves from both sides of the end-point did not give the true end-point but one which, if the technique were standardized, was very reproducible. In this way silver(I) and mercury(II) were determined in the concentration range 0·1–1000 mg/kg with a precision of $\pm0\cdot3\%$ for 100–1000 mg/kg and up to $\pm1\%$ for the extremely low concentrations between 0·1 and 100 mg/kg. The potassium iodide titrant used had a concentration of between 0·1 and 10 mM according to the level of cation concentration.

3.52.2 *Manganese(II) catalyst*

Two of the most stable complexes of manganese(II) are those formed with EDTA and with DCTA and it follows therefore that these two chelating compounds can be determined by titration with manganese(II), and also that if an excess of the chelating agent is present after complex formation with another cation the excess can be determined in a similar way. This back-titration has been used to determine small amounts of

**Table 3.84 Determination of EDTA and DCTA by the
catalytic indicator method**
(From Weisz and Kiss (1970))

Substance	Taken	Hydrogen peroxide-resorcinol		Hydrogen peroxide	
		Found	*Standard deviation*	*Found*	*Standard deviation*
	μg	μg	$\pm\mu$g	μg	$\pm\mu$g
EDTA	2900	2941	21·3	2895	20·4
	290·1	292·2	1·82	293·1	1·90
	14·51	14·63	0·19	14·58	0·15
	1·450	1·472	0·01	1·445	0·02
DCTA	3463	3460	22·3	3408	21·4
	346·3	340·2	1·60	341·6	1·52
	17·31	17·26	0·16	17·42	0·18
	1·731	1·714	0·02	1·682	0·02

twelve different metal ions (Weisz and Kiss, 1970). It is also possible to determine traces of manganese(II) by titrating the test solution into an EDTA solution containing the system that can be catalysed.

In the general method, 0·5 ml of test solution was pipetted into a 10-ml Dewar followed by 1·5 ml of water, 0·5 ml of 1 M ammonium carbonate, 0·1 ml of 5% hydrogen peroxide (plus 0·1 ml of 5% resorcinol solution if required) and the mixture was titrated with 0·02 M or 0·0001 M $MnCl_2$ at a rate of approximately 50 μl/min.

Table 3.84 shows the results obtained in the determination of EDTA and

DCTA by direct titration and Table 3.85 those obtained in the determination of metal ions by the back-titration method. The precision of the tests given in the tables shows that better results are obtained with the hydrogen peroxide decomposition reaction.

Table 3.85 Determination of small amounts of metal ions by the catalytic indicator method
(From Weisz and Kiss (1970))

Cation Determined	*Taken* μg	*Hydrogen peroxide–resorcinol*		*Hydrogen peroxide*	
		Found μg	*Standard deviation* μg	*Found* μg	*Standard deviation* μg
Zn^{2+}	653·8	659·9	0·95	648·2	1·01
	130·8	129·6	0·43	132·4	0·67
	65·38	65·95	0·34	65·01	0·43
Cd^{2+}	1124	1135	12·4	1116	10·8
	224·8	222·7	0·23	226·9	0·76
	112·4	114·5	0·30	110·3	0·32
Cu^{2+}	635·4	630·0	0·85	639·5	0·78
	127·1	125·8	0·25	128·1	0·35
	63·54	63·98	0·18	63·01	0·28
Ni^{2+}	587·1	589·7	0·89	582·5	0·13
	117·4	119·5	0·55	115·4	0·64
	58·71	58·20	0·27	59·21	0·35
Mn^{2+}	549·4	543·7	0·67	554·8	0·29
	109·9	111·4	0·34	108·2	0·35
	54·94	54·01	0·15	54·90	0·17
Pb^{2+}	2072	2093	11·3	2046	9·87
	414·4	418·2	0·84	410·3	0·78
	207·2	204·4	0·32	209·9	0·43
Hg^{2+}	2006	2021	12·4	2001	10·3
	401·2	405·9	0·87	400·0	0·67
	200·6	204·5	0·54	196·2	0·46
Sn^{2+}	1187	1195	10·3	1170	11·0
	237·4	237·0	0·34	234·2	0·42
	118·7	119·5	0·12	119·3	0·20
Al^{3+}	269·8	267·2	0·44	272·3	0·38
	53·96	53·32	0·14	53·40	0·19
	26·98	26·51	0·08	27·12	0·10
In^{3+}	1148	1152	10·1	1132	9·8
	229·6	225·3	0·76	232·0	0·76
	114·8	116·0	0·12	112·2	0·18
Ga^{3+}	697·2	690·0	0·86	691·1	0·98
	139·4	141·1	0·32	137·2	0·30
	69·72	69·10	0·22	70·71	0·24
Th^{4+}	2320	2345	12·2	2301	11·4
	464·0	468·2	0·81	460·3	0·91
	232·0	230·4	0·41	234·2	0·38

3.52.3 *Displacement titrations*

Kiss (1970) has extended the principle of the catalytic end-point method. He added to the titrand a substance that contained the bound catalytic substance, and the titrant used was one that would displace the bound substance to form the ion which was to act as the catalyst. For example, thiocyanate ions used as a titrant will displace iodide from insoluble silver(I), mercury(II) or palladium(II) iodides and this iodide then catalyses the As(III)–Ce(IV) system as described in Section 3.52.1 above. Substances that react with thiocyanate ions, such as silver(I), mercury(II) or palladium(II) can therefore be titrated at low concentrations. In the method, about 0·2 mg of one of these iodides was added to the titrand and 20 mM KSCN was used as the titrant in a titration system similar to that outlined in Section 3.35.1.

The stable manganese(II)–EDTA complex, when added to the titrand, enables a number of metal ions (given in Table 3.85) to be used as titrants in the determination of traces of EDTA. The liberated manganese(II) ions act as the catalyst in the decomposition of hydrogen peroxide as described in Section 3.35.2 above.

Chapter 4

Organic Reactions

Because of the importance of acidic and basic organic substances both as pure substances and as impurities, e.g., as acidity in other organic substances, this chapter deals first with neutralization reactions and then with other organic reactions, listed alphabetically as the functional group activity which is being investigated or determined.

Organic substances are predominantly insoluble in water, so most of their reactions have to be studied in non-aqueous systems and consequently care has to be taken to avoid extraneous heat effects, particularly those due to heats of mixing and heat loss by evaporation if the solvent is volatile. Organic substances containing metal radicals are covered in this chapter if their reactions and properties are concerned primarily with the organic radical, otherwise they are discussed in Chapter 3.

4.1 Neutralization reactions

In neutralization reactions involving organic substances there is a choice between using water or an organic substance as the solvent in the titration, depending on the solubility of the substances being titrated, but many more organic solvents can be used than is possible in the neutralization of inorganic substances. The value of the dielectric constant of the solvent is important when making the choice. If the acid to be titrated is not charged, the value of the dielectric constant is not important, but if the substance is a cationic acid or an anionic base it is preferable to use a solvent having a low dielectric constant, and if the substance is an anionic acid or a cationic base a solvent having a high dielectric constant should preferably be used.

The acidity or basicity of an amphiprotic solvent is also an important factor. Weak bases have their basic properties enhanced in an acidic solvent, and in a basic solvent the acidic properties of weak acids are enhanced. These increases in strength are accompanied by a 'levelling' process that brings the individual strengths of acids (or bases) closer to each other so that components of mixtures of acids or bases are less likely to be distinguished from one another. If components of a mixture are to be distinguished it is better to use an aprotic solvent. These solvents enhance the differences in acidic (and basic) strengths between acids (and bases) having the same charge type, and the change in hydrogen ion concentration that takes place on neutralization is then much larger than in amphiprotic solvents. The dielectric constants of common solvents, together with the autoprotolytic constants of some of the amphiprotic solvents, are given in Table 4.1 as a guide to those that may be used.

Table 4.1 Properties of common solvents

| | Amphiprotic solvents | | Aprotic solvents | |
Solvent	Dielectric constant	Autoprotolytic constant	Solvent	Dielectric constant
Water	78·5	14·0	Dimethylsulphoxide	45·0
Formic acid	58·5	6·2	Acetonitrile	36·0
Ethylene glycol	37·7	—	Nitrobenzene	34·9
Methanol	32·6	16·7	Acetone	20·7
Ethanol	24·3	19·1	Tetrahydrofuran	7·6
Ethylenediamine	12·9	—	Diethyl ether	4·3
Pyridine	12·3	—	Benzene	2·3
Acetic acid	6·1	14·5	Dioxan	2·2

Hansen and Lewis (1972) have proved their 'total moles–total enthalpy' method by titrating (*i*) a mixture of glycine ($\Delta H = -2\cdot80$ kcal/mole, $pK = 9\cdot8$) and phenol ($\Delta H = -7\cdot75$ kcal/mole, $pK = 11$) with alkali and (*ii*) a mixture of sodium acetate ($\Delta H = 0\cdot5$ kcal/mole, $pK = 4\cdot8$) and pyridine ($\Delta H = -4\cdot95$ kcal/mole, $pK = 5\cdot3$) with acid. They obtained a relative error of about 5% at the millimolar level, thus opening a new approach to the analysis of mixtures of acids (or bases) with equal pK values.

In neutralization reactions involving organic acids or bases in aqueous solution, similar arguments apply to the conditions of the titration as those given under the neutralization of inorganic substances (see Section 3.48). Thus it is no accident that the earliest recorded thermometric titration was developed because of the difficulty in preparing a neutral solution of ammonium citrate. Bell and Cowell (1913) noticed that a large quantity of heat was developed in this neutralization and that none was produced after the neutralization was judged to be complete, and it was from this observation that the early titration method was developed and put on a truly analytical basis by Dutoit and Grobet in 1922.

In order to standardize the titrant in neutralization reactions in non-aqueous solution it is necessary to use pure, readily available substances, and the author has used benzoic acid (see Table 4.2) in the titration of acids, and 8-hydroxyquinoline in the titration of bases.

4.1.1 Acids in aqueous solution

(*i*) MONOPROTIC CARBOXYLIC ACIDS

Monoprotic carboxylic acids have been titrated with sodium hydroxide in aqueous solution by a number of workers and the conditions used and the precision of the results obtained are given in Table 4.2.

Barthel and his co-workers (1961) titrated the individual components of mixtures of hydrochloric acid with acetic acid, propionic acid or monochloroacetic acid and obtained a precision of $\pm0.5\%$ for the hydrochloric acid end-point and $\pm0.2\%$ for the second, organic acid end-point.

The thermodynamic data for the neutralization and ionization of acetic acid and its halogen-substituted derivatives were determined by Jordan

Table 4.2 Conditions and precision of results in the titration of monoprotic carboxylic acids in aqueous solution

Acid	Taken	Volume ml	Titrant NaOH, M	Precision $\pm\%$	Apparatus	Ref.
Acetic*	68 meq	60	0·98	1·0	continuous-flow burette; thermistor	(a)
Acetic	0·05–0·50 arbitrary units	—	0·6	2	continuous-flow titrator	(b)
Acetic*	3–100 mM	50	1·0	1·0	syringe burette; thermistor	(c)
Benzoic	0·3 mM	50	2·0	0·65	syringe burette; thermistor	(d)
Salicylic	0·3 mM	50	2·0	1·2	syringe burette; thermistor	(d)

* Similar results obtained with monochloro- and trichloroacetic acids (a) Linde *et al.* (1953); (b) Priestley *et al.* (1965); (c) Jordan and Dumbaugh (1959a); (d) Harries (1968)

and Dumbaugh (1959a, b) and also by Dumbaugh (1959); Avedikian (1966) obtained further data on the same subject and also studied propionic acid and its halogen derivatives. Jordan and Dumbaugh used the conditions given in Table 4.2, Avedikian used a Titra-Thermo-Mat apparatus, titrating 20 ml of a 5 mM solution of the acid with 2M NaOH, and Barthel and his co-workers (1961) the early titration method. The data given in Table 4.3 include those obtained for monosodium tartrate acting as a monoprotic acid, and the striking regularity of the differences between the enthalpies of ionization is evident.

Jordan (1963b) remarked that a significant contribution to important theoretical studies resulted from thermochemical titrations and exemplified this in relation to the entropies of ionization of the halo-acetic acids. If the entropies determined for the 0, 1, 2 and 3 halogen-substituted acids are noted, they can be seen to differ for each increase of one halogen substituent by a value of 6–7 cal mole^{-1} degK^{-1}. That this value is similar to the molar entropy of the freezing of water means that the differences between the numbers of water molecules associated with the unionized and ionized acid are 3, 2, 1 and 0, according to the general equation:

$$CH_{3-x}Hal_xCOOH + (4-x)wH_2O$$
$$\rightleftharpoons CH_{3-x}Hal_x - COO^-(w+3-x)H_2O + H_3O^+$$

where $x = 0, 1, 2,$ or 3. It can be seen from the equation that the difference for each value of x is one molecule of water removed from a 'frozen' state.

Table 4.3 Thermodynamic data for the neutralization and ionization of monoprotic carboxylic acids in aqueous solution

Acid	*Neutralization* $-\Delta H^\circ$ kcal/mole	$-\Delta H^\circ$ kcal/mole	*Ionization* $-\Delta G^\circ$ kcal/mole	$-\Delta S^\circ$ cal mole^{-1} degK^{-1}	*Ref.*
CH_3COOH	$13\cdot50 \pm 0\cdot40$	$0\cdot0 \pm 0\cdot3$	$6\cdot6$	22 ± 1	(a)
	$13\cdot61 \pm 0\cdot05$	$0\cdot09$	$6\cdot49$	$22\cdot1$	(b)
CH_2FCOOH	$14\cdot20 \pm 0\cdot10$	$1\cdot0 \pm 0\cdot1$	$3\cdot5$	15 ± 1	(a)
	$14\cdot20 \pm 0\cdot04$	$0\cdot68$	$3\cdot53$	$14\cdot1$	(b)
$CH_2ClCOOH$	$14\cdot40 \pm 0\cdot10$	$1\cdot3 \pm 0\cdot1$	$4\cdot1$	18 ± 1	(a)
	$14\cdot25 \pm 0\cdot4$	$0\cdot3$	$3\cdot91$	$15\cdot6$	(b)
	$14\cdot47$	—	—	—	(a)
$CH_2BrCOOH$	$13\cdot90 \pm 0\cdot2$	$0\cdot5 \pm 0\cdot3$	$4\cdot0$	15 ± 1	(a)
	$14\cdot14 \pm 0\cdot06$	$0\cdot2$	$3\cdot96$	$15\cdot4$	(b)
CH_2ICOOH	$14\cdot10 \pm 0\cdot1$	$0\cdot8 \pm 0\cdot1$	$4\cdot3$	17 ± 1	(a)
	$14\cdot68 \pm 0\cdot05$	$1\cdot16$	$4\cdot33$	$18\cdot4$	(b)
CHF_2COOH	$13\cdot50 \pm 0\cdot2$	$0\cdot0 \pm 0\cdot8$	$1\cdot8$	6 ± 3	(a)
$CHCl_2COOH$	$13\cdot60 \pm 0\cdot1$	$0\cdot1 \pm 0\cdot5$	$1\cdot7$	6 ± 2	(a)
	$13\cdot69 \pm 0\cdot05$	$0\cdot17$	$1\cdot77$	$6\cdot5$	(b)
$CHBr_2COOH$	$13\cdot60 \pm 0\cdot2$	$0\cdot5 \pm 0\cdot9$	$1\cdot9$	8 ± 3	(a)
CF_3COOH	$13\cdot50 \pm 0\cdot1$	$0\cdot0 \pm 0\cdot9$	$0\cdot3$	1 ± 3	(a)
CCl_3COOH	$13\cdot40 \pm 0\cdot1$	$-1\cdot5 \pm 1\cdot5$	$2\cdot1$	-2 ± 5	(a)
CBr_3COOH	$13\cdot20 \pm 0\cdot1$	$0\cdot8 \pm 0\cdot3$	$-0\cdot2$	2 ± 1	(a)
CH_3CH_2COOH	$13\cdot70 \pm 0\cdot05$	$0\cdot18$	$6\cdot65$	$22\cdot9$	(b)
$CH_3CHClCOOH$	$14\cdot46 \pm 0\cdot05$	$0\cdot94$	$3\cdot86$	$16\cdot1$	(b)
$CH_3CHBrCOOH$	$14\cdot80 \pm 0\cdot07$	$1\cdot28$	$4\cdot50$	$17\cdot9$	(b)
CH_2ClCH_2COOH	$14\cdot30 \pm 0\cdot06$	$0\cdot78$	$5\cdot54$	$21\cdot2$	(b)
CH_2BrCH_2COOH	$15\cdot24 \pm 0\cdot07$	$1\cdot72$	$5\cdot41$	$23\cdot9$	(b)
CH_2ICH_2COOH	$14\cdot90 \pm 0\cdot05$	$1\cdot38$	$5\cdot54$	$23\cdot2$	(b)
CH_3CCl_2COOH	$13\cdot97 \pm 0\cdot04$	$0\cdot45$	$2\cdot80$	$10\cdot9$	(b)
$CH_2ClCHClCOOH$	$13\cdot68 \pm 0\cdot05$	$0\cdot17$	$3\cdot87$	$13\cdot6$	(b)
$CH_2BrCHBrCOOH$	$15\cdot41 \pm 0\cdot08$	$1\cdot89$	$3\cdot17$	$17\cdot0$	(b)
$NaOOCCHOH$ $CHOHCOOH$	$13\cdot60 \pm 0\cdot3$	$0\cdot1$	—	20 ± 1	(a)
$CH_3(CH_2)_2COOH$	$14\cdot05$	—	—	—	(c)

(a) Jordan and Dumbaugh (1959a, b); Dumbaugh (1959)
(b) Avedikian (1966)
(c) Barthel *et al.* (1961)

(*ii*) DIPROTIC CARBOXYLIC ACIDS

A number of diprotic carboxylic acids have been titrated by Harries in 50 ml of solution with 2 M NaOH. The details and the precision of his results are given in Table 4.4. One of the most interesting facts to emerge from his work was that the two ionization stages of succinic acid, with a pK difference as small as $1\cdot45$, could be distinguished in the titration.

Those acids in which the two ionization stages could be distinguished showed a steeper titration graph for the second ionization than the first except in the case of phthalic acid. This is a reflection of the heats of ionization of the two stages which Harries also determined and which are

Table 4.4 Precision of results in the titration of diprotic carboxylic acids
(From Harries (1968))

Acid	Ratio found/added	End-point ratio 1st/2nd	Relative standard deviation $\pm\%$	No. of tests
Oxalic	0·976	1·88	2·0	5
Malonic	1·008	2·00	0·5	6
Succinic	1·001	1·97	2·2	5
Glutaric	1·018	(a)	0·65	2
Adipic	0·982	(a)	2·1	3
Maleic	1·010	(a)	1·4	5
Fumaric	0·995	(a)	0·7	2
Tartaric	1·022	(a)	1·6	3
Phthalic	0·991	1·98	0·9	7

(a) No intermediate half-neutralization end-points.

Table 4.5 Thermodynamic data for the neutralization and ionization of diprotic carboxylic acids in aqueous solution

Acid	Neutralization $-\Delta H°$ kcal/mole	Ionization		Ref.
		$-\Delta H°$ kcal/mole	$-\Delta S°$ cal mole^{-1} degK^{-1}	
Oxalic	27·7 $\pm$ 0·2	0·7 $\pm$ 0·7	32 $\pm$ 2	(a)
	13·5*			
	14·6†	—	—	(b)
Malonic	27·9 $\pm$ 0·6	−0·9 $\pm$ 9·7	39 $\pm$ 2	(a)
	14·6*			
	15·1†	—	—	(b)
Succinic	28·5 $\pm$ 0·6	−1·5 $\pm$ 0·6	38 $\pm$ 2	(a)
	13·2*			
	14·0†	—	—	(b)
Glutaric	28·3 $\pm$ 0·2	1·3 $\pm$ 0·2	48 $\pm$ 1	(a)
	28·6	—	—	(b)
Adipic	28·7 $\pm$ 0·4	1·7 $\pm$ 0·4	50 $\pm$ 1	(a)
	28·8	—	—	(b)
Pimelic	28·6 $\pm$ 0·3	1·6 $\pm$ 0·3	50 $\pm$ 1	(a)
Azelaic	28·5 $\pm$ 0·3	1·5 $\pm$ 0·3	50 $\pm$ 1	(a)
Tartaric	26·9 $\pm$ 0·4	−0·1 $\pm$ 0·3	34 $\pm$ 2	(a)
	27·3	—	—	(b)
Maleic	28·4	—	—	(b)
Fumaric	28·5	—	—	(b)
Phthalic	14·4*	—	—	(b)
	13·7†			

* First ionization; † Second ionization
(a) Jordan and Dumbaugh (1959b); (b) Harries (1968)

shown in Table 4.5 together with the results obtained by Jordan and Dumbaugh (1959b) for the thermodynamic data of both the neutralization and ionization stages of diprotic acids under the conditions given in Table 4.2.

In the series from oxalic to azelaic acid the entropy of ionization becomes increasingly negative to a minimum of approximately -50 cal/mole^{-1} degK^{-1} which implies that the length of the chain of the molecule and thus the separation of the charges on the carboxylic acid groups is the cause of this.

Priestley and his co-workers (1968) reported that they had titrated tartaric acid, using a continuous-flow titrator.

(*iii*) TRIPROTIC CARBOXYLIC ACIDS

The neutralization of citric acid with ammonia was the first reported example of a thermometric titration (Bell and Cowell, 1913). Using modern apparatus Barthel and his co-workers (1961) titrated the acid and obtained results that had a precision of $\pm 5\%$ whilst Jordan and Dumbaugh (1959b) determined the enthalpy of neutralization, $\Delta H°$, as $-41·0 \pm 0·04$ kcal/mole, and the enthalpy, $\Delta H°$, and entropy, $\Delta S°$, of ionization as $-0·6 \pm 0·4$ kcal/mole and 67 ± 1 cal mole^{-1} degK^{-1} respectively. Only one end-point was observed in these aqueous titrations of this triprotic carboxylic acid.

(*iv*) AMINO-ACIDS

The first investigation into the titration of amino-acids was undertaken by Chatterji and Ghosh (1957), using the early titration method, and it was shown that glycine and alanine could be titrated with one sharp end-point whereas the diprotic acids glutamic acid and aspartic acid both showed two end-points, indicating that both ionizable hydrogen atoms could be titrated. The solutions of the acids were approximately 0·1 N and 0·5–2·0 N NaOH was used as the titrant. The error of the method was estimated as $\pm 1·2\%$. Similar results were obtained by Sen and Wu (1969) using much lower concentrations of the amino-acids (0·1–1·0 mM) and a sodium hydroxide titrant of such concentration that the total titration volume was about 1 ml.

These methods are decidedly more elegant than other more involved and time-consuming procedures, e.g., titration before and after adding formaldehyde or the manometric methods using specific enzymes.

All the titration graphs of monoprotic acids showed a single sharp inflection at the end-point and all those of the diprotic amino-acids showed two, the first part of the graph always being very steep. They showed that it was possible to titrate mixtures of mono- and diprotic amino-acids. In this case the first inflection represented the first ionization stage of the diprotic acid and the second inflection represented the second ionization stage of the diprotic acid together with the only ionization stage of the monoprotic acid. For the neutral amino-acids the titration results became unreliable (error greater than $\pm 5\%$) at concentrations of less than 6 mM whereas the acidic diprotic amino-acids gave good results even at the 0·1 mM level. Sen and Wu combined this work with an NMR examination and stated that the combination of the two techniques proved an elegant method for the identification of amino-acids and their binary mixtures.

Avedikian (1967) determined the enthalpy of neutralization and the free energy and entropy of ionization of eleven amino-acids, and the results are given in Table 4.6.

Table 4.6 Thermodynamic data for the neutralization and ionization of amino-acids
(From Avedikian (1967))

Amino-acid R—COOH	*Neutralization* $-\Delta H^0$ kcal/mole	*Ionization* $-\Delta G^0$ kcal/mole	 $-\Delta S^0$ cal mole^{-1} degK^{-1}
$NH_2\ CH_2$—	2·88	13·29	8·90
$NH_2\ (CH_2)_2$—	2·50	13·90	8·82
$NH_2\ (CH_2)_3$—	1·57	13·35	8·05
$NH_2\ (CH_2)_4$—	0·56	14·63	5·57
$CH_3\ CH(NH_2)$—	2·77	13·41	8·93
$CH_3\ CH_2\ CH(NH_2)$—	2·68	13·36	8·32
$CH_3\ (CH_2)_2\ CH(NH_2)$—	2·67	13·33	8·46
$(CH_3)_2\ CH(NH_2)$—	2·43	13·86	8·29
$(CH_3)_2\ CH\ CH(NH_2)$—	2·88	13·24	8·72
$CH_3\ NH\ CH_2$—	3·78	13·83	13·72
$CH_3\ CO\ NH\ CH_2$—	3·78	5·00	17·48

The enthalpy and pK values at 25 °C over the range of ionic strengths from 0·002 to 1·0 were determined for the proton ionization from (*i*) the NH_3^+ and S–H groups of L-cysteine, (*ii*) the NH_3^+ group of *S*-methyl-L-cysteine and (*iii*) the S–H group of mercaptoacetic acid (Wrathall *et al.*, 1964). The heats of neutralization, dilution and solution of glutamic acid and of its chlorohydrate and glycoside were determined by Goudard and Grangetto (1966); for details of the studies of the association of copper(II) with amino-acids see Sections 3.30.7.

The velocity constants of the reaction of carbon dioxide with the anions of the amino-acids, glycine, glycylglycine, α-alanine, β-alanine, valine, histidine, β-phenylalanine, α-aminophenylacetic acid, and ε-amino-caproic acid have been determined by the rapid thermal method of Roughton (Chipperfield, 1966).

(*v*) PHENOLS

The lower monohydric phenols and the polyhydric phenols are soluble in water. For this reason most of the titrations of monohydric phenols have been investigated in non-aqueous solution, but in spite of the statement in a recent review that it is impossible to titrate phenols in aqueous solution with a strong base because of their weak acidity (the pK of *o*-cresol is 10·29), this was achieved by Pandya and Haldar (1962), using the early titration method. They investigated the titration of phenol, *o*-cresol, *m*-nitrophenol and resorcinol in 50 ml of solution with 0·25 M

NaOH as titrant. All the phenols showed sharp end-points at concentrations between 1·8 and 6·4 g/l and the error was found to be less than $\pm1\%$. For resorcinol concentrations above 0·025 M the titration graph showed two inflections representing the successive titration of the two hydroxyl groups ($T = 1·25$ degC/mmole). The method is not affected by the presence of amines and alcohols and has been applied to concentrations as low as 0·01 M (0·9 g/l) by the author (see Fig. 1.4).

In the titration of phenol with alkali an error of less than $\pm0·7\%$ and an enthalpy of neutralization of $-7·80 \pm 0·10$ kcal/mole has been obtained by using a 'corrected non-adiabatic' system (Nakanishi and Fujieda, 1972).

Aspirin in tablets has been titrated with alkali and a result of 295 mg was obtained compared with the expected 293 mg (Beezer, 1970).

(*vi*) Pyrimidines, purines, nucleotides and nucleosides

Christensen and his co-workers have shown that it is possible to study the deprotonation of a number of important biological materials. Their determinations were carried out in dilute aqueous solution with 0·06 M NaOH as the titrant, and the standard enthalpy, entropy and free-energy changes were calculated (cf. Hansen *et al.*, 1965; Christensen *et al.*, 1966b) for the loss of one proton. These values were calculated (Christensen and Izatt, 1962) for the pyrimidine-NH (pK 4), phosphate (pK 7 and pK 9) and the imidazole-NH (pK 9) groups present in adenosine and its

Table 4.7 Values of pK and $\Delta H°$ for the deprotonation of nucleosides

Compound	Reaction	pK	$-\Delta H°$ kcal/mole	Method
Adenosine(ad)	ad = H$^+$ + ad$^-$	12·35 $\pm$ 0·03	9·7 $\pm$ 0·2	(*a*)
		12·5		(*b*)
Cytosine(cy)	cy = H$^+$ + cy$^-$	12·15 $\pm$ 0·05	11·5 $\pm$ 0·1	(*c*)
		12·16		(*b*)

(*a*) Thermometric;   Izatt *et al.* (1966)
(*b*) Potentiometric;   Levene *et al.* (1926)
(*c*) Thermometric;   Christensen *et al.* (1967)

mono-, di- and tri-phosphates. They were also determined for adenine, adenosine, 2-deoxyadenine, 3-deoxyadenosine, 2-*O*-methyladenosine, adenosine-5′-monophosphate, adenosine-3′, 5′-cyclophosphate, adenosine-2′(3′)-monophosphate, the associated sugars ribose, 2-deoxyribose, arabinose and fructose (Christensen and Izatt, 1962; Izatt *et al.*, 1965; 1966) and for 9-β-D-xylofuranosyladenine (Christensen *et al.*, 1966b).

The pK values of the sugars and the adenosine and its 5′-monophosphate were found to be between 12 and 13 which shows the great superiority of the thermometric method compared to the potentiometric method. From these findings it was shown that the acidity of the adenosine was associated with its ribose moiety.

A potentiometric titration was used to determine pK values in the range 4·0–10·0 whereas a thermometric titration was preferred for the determination of both the pK values in the range 11·5–13·5 and the enthalpies of deprotonation of cytosine, cytidine, thymine, thymidine, uracil and uridine. The pK for the second ionization of thymidine (12·85) was the highest found and was attributed to the presence of the deoxyribose moiety. The results obtained showed that the increased negative charge resulting from ionization of the nucleoside did not affect the second ionization, indicating that they are well removed from each other (Christensen *et al.*, 1967). Some of the results of this investigation are given in Table 4.7 and are compared with those obtained potentiometrically to show the good agreement obtained for the pK values.

(*vii*) PROTEINS

Recent work by Jespersen and Jordan (1970) showed that the carboxyl, imidazole and amine prototropic groups in proteins can be titrated separately in a single thermometric titration with sodium hydroxide, whereas a potentiometric titration determines only the carboxyl groups.

Chipperfield and his co-workers (1967) used the rapid-flow calorimeter to determine the heats of ionization of oxygenated and deoxygenated haemoglobins.

(*viii*) SOIL

Ragland (1962) used thermometric titrimetry to study the cation-exchange capacity and heat of neutralization of clays. The sample, 0·5 g of bentonite or 0·1 g of peat, was suspended in 50 ml of water and titrated with 0·45 N NaOH. The mesh-size of the soil was found to be important; 20-mesh gave a very rounded end-point, 200-mesh was satisfactory though the end-point was still rounded. The H^+-saturated bentonites gave steeper titration curves than the Al-saturated bentonites and mixtures of the two types gave an inflected titration graph, the H^+-type being titrated first, thereby enabling the two types of clays to be distinguished.

(*ix*) WEAK-BASE SALTS

As in the titration of inorganic salt, it is possible to titrate weak organic base salts with a strong base, using the thermometric technique. This

Table 4.8 Titration of weak-base salts in pharmaceutical analysis
(From De Leo and Stern (1966))

Substance	Taken M	No. of tests	Titrant NaOH	Found %
Chlorpheniramine maleate	0·01	5	2·0 M	97·8 ± 0·6
Chlorpromazine hydrochloride	0·01	5	1·0 M	99·7 ± 0·3
	0·01	6	2·0 M	98·6 ± 0·3
Hydrochlorothiazide	0·0027	5	1·0 M	99·7 ± 0·2

'displacement' reaction has been used in the analysis of pharmaceuticals by De Leo and Stern (1964, 1966). The substances were titrated in 100 ml of solution and the results obtained are given in Table 4.8.

Chloropromazine hydrochloride was determined in single tablets containing the drug and the results were found to be 207·0 ± 1·9 mg, well within the 210·4 ± 2·9 mg expected in each tablet. The great advantage of the thermometric method compared to the U.S.P. potentiometric perchloric acid titration in the presence of mercury(II) acetate is that in the latter method both the chlorpheniramine and maleate are titrated together whereas in the thermometric method they are titrated separately, the titration graph showing two distinct end-points.

Beezer (1971) has titrated codeine phosphate with alkali. The tablets containing the drug were dissolved in 5% water–ethanol and titrated

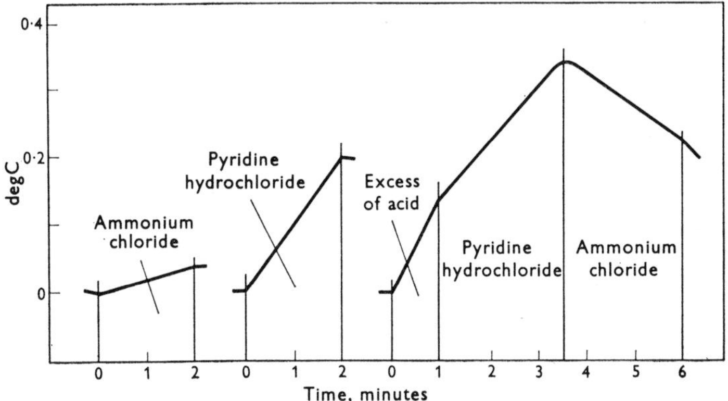

Fig. 4.1 Graphs of the enthalpimetric titration of aqueous solutions of nitrogen bases with 5M NaOH; from Vaughan and Swithenbank (1967).

immediately. A result of 29·07% was obtained compared with 28·2% obtained by the British Pharmacopoeia method.

If a mixture of a weak base, for example ammonia or methylamine, and of a weaker base, for example pyridine, is acidified and then titrated with a strong base, the titration graph shows first the neutralization of the excess of acid added, with a very sharp temperature rise, second, the titration of the weaker base salt with a less sharp temperature rise, and finally the titration of the weak base, with a barely discernible temperature rise (Parsons, 1957; Vaughan and Swithenbank, 1967). Figure 4.1 shows the titration of (i) 0·1 M NH_4Cl (see also Section 3.48.1 (ix)), (ii) 0·1 M pyridine hydrochloride and (iii) an equimolar solution of the two bases. The titration volume in each case was 5 ml and 5 M NaOH was used as the titrant.

Vaughan and Swithenbank (1967) applied the method to the analysis of tar-plant aqueous liquors for their pyridine-base contents, as in many cases the high ammonia content precluded the use of a direct titration because of the large amount of heat evolved. Some of the results obtained are given in Table 4.9 and, where possible, are compared with other methods. It can also be seen from the table that 0·3% of pyridine can be determined in the presence of ten times as much ammonia, a determination that cannot be carried out with such ease by any other method.

This method was also applied to the determination of the free sulphuric acid and pyridine-base content of sulphuric acid extracts of tar-oils containing bases (Vaughan, 1966).

In the method 1 ml of sample was diluted to 5 ml before being titrated with 5 M NaOH. The results obtained showed that pyridine bases at the

Table 4.9 Pyridine base content of tar-plant aqueous liquors after acidification
(From Vaughan and Swithenbank (1967))

Source	*Basic nitrogen content % as pyridine*
Distilled from phenol rectifiers	
a	0·60, 0·63 (0·51*)
b	0·33, 0·35†
c	0·11, 0·11 (0·14*)
Ammonium sulphate liquors	
d	0·56, 0·56 (0·51‡)
e	9·2, 9·4

* Acid titration after ammonia removal with formalin
† Contained 2·5% of ammonia
‡ Perchloric acid titration of toluene extract of alkaline distillate

30–40% level could be determined to within ±0·2% in the presence of 1–20% of sulphuric acid which itself could be determined with the same precision.

With a more complex mixture of bases, Vaughan and Swithenbank (1967) showed that pyridine, morpholine and ethanolamine could be titrated separately in admixture in a single titration after acidification. The titration graph obtained in this titration is shown in Fig. 1.6.

The very small differences in the thermodynamic properties of diastereoisomeric compounds are difficult to determine but it has been shown that it is possible to use thermometric titrimetry for this purpose. Two compounds of this type have been studied: ephedrine and pseudoephedrine (Raffa *et al.*, 1968). Either the substances were titrated with 1·0 M HCl or their hydrochlorides were titrated with 1·0 M NaOH and the results of eight or nine replicate determinations of five concentrations in the range 1–10 mM were treated statistically and the results obtained are given in Table 4.10, in which they are compared with results obtained potentiometrically.

Basic amino-acids have been titrated as their hydrochlorides by Sen and Wu (1969). The titrant used was sodium hydroxide, but the authors do not say whether any free hydrochloric acid present in the hydrochlorides would interfere. In the titration of 1-α-histidine hydrochloride quantities of between 24 and 100 mg (20–93 mM) were determined with an error of 1·0–1·5% and a precision of better than ±0·13%. See also Section 4.1.1(*iv*) for the titration of these compounds as acids.

Table 4.10 Thermodynamic data for the dissociation of the diastereoisomeric ephedrinium ions at 25 °C

	$\Delta H°$ cal/mole	$\Delta G°$ cal/mole	$-\Delta S°$ cal/mole^{-1} degK^{-1}	*Method*
Ephedrine	10·84 ± 0·04	12·80 ± 0·02	6·57 ± 0·13	thermometric
hydrochloride	—	13·05 ± 0·01	7·42 ± 0·12	potentiometric*
(A)	10·83 ± 0·04	13·02 ± 0·01	7·35 ± 0·25	potentiometric†
Pseudoephedrine	11·04 ± 0·02	13·00 ± 0·01	6·57 ± 0·24	thermometric
hydrochloride	—	13·25 ± 0·01	7·41 ± 0·06	potentiometric*
(B)	11·03 ± 0·04	13·25 ± 0·01	7·42 ± 0·25	potentiometric†
Difference (B–A)	0·20 ± 0·04	0·20 ± 0·07	0·00 ± 0·27	thermometric
	—	0·20 ± 0·01	0·01 ± 0·13	potentiometric*
	0·20 ± 0·04	0·23 ± 0·01	0·07 ± 0·35	potentiometric†

*Raffa *et al.* (1968);   † Everett and Hyne (1958)

4.1.2 *Acids in non-aqueous solution—normal end-point methods*

(*i*) CARBOXYLIC ACIDS

The solvent used in the non-aqueous titration of acids is important in that it affects the choice of base that can be used for the titration, as it is preferable to use the same solvent for both the substance and the titrant. Harries (1968) used 1·5 M sodium methoxide and also 0·5 M tetra-n-butylammonium hydroxide, both in methanol solution; with the latter titrant the sample was dissolved in a 1 : 4 methanol–benzene mixture which gave no heat of dilution effects with the titrant. Table 4.11 gives the results obtained by using the two titrants, and if these results are compared with the results in the aqueous titration for the diprotic acids given in Table 4.4 it can be seen that in the non-aqueous system both ionization stages of maleic, fumaric and tartaric acids are titratable whereas in the aqueous titration they are not.

The titration graphs rose more steeply after the first neutralization in the case of phthalic, malonic and succinic acids with both titrant–solvent systems, and this also occurred in the case of oxalic, glutaric and tartaric acids but only with the quaternary base titrant system. This and the opposite behaviour of glutaric and adipic acids with the two titrants highlights the differences that occur in the neutralization enthalpy, the overall values of which are given in Table 4.12.

Acetonitrile, dried with molecular sieve 4A, has also been used as a solvent with 1,3-diphenylguanidine as the titrant (Forman and Hume, 1964). The analytical results and thermodynamic data obtained are given in Table 4.13. In this titration system both acetic and *p*-toluic acids were found to be too weak to be titrated and *m*-nitrobenzoic acid was found to have an abnormally high heat of neutralization.

A direct-injection enthalpimetric method was investigated in a search for a rapid method for the determination of acetic, propionic and benzoic acids in acetic anhydride. In this method $1 \cdot 2 \pm 0 \cdot 1$ ml of triethylamine was injected into 20 ml of sample in a small Dewar flask and the temperature

Table 4.11 Titration of carboxylic acids in non-aqueous solution
(From Harries (1968))

Acid	Ratio found/added	End-point ratio 1st/2nd	Relative standard deviation $\pm\%$	No. of tests	$-\Delta H°$ kcal/mole	Method
Benzoic	1·000	—	0·83	8	7·0	A
	1·000	—	0·55	8	5·9	B
Salicylic	0·992	—	0·42	3	7·1	A
	1·012	—	0·67	4	5·7	B
Phthalic	0·998	1·99	0·83	5	6·5	A
	1·013	1·99	0·50	5	4·4	B
Isophthalic	1·013	2·08	0·90	4	7·2	A
	0·967	1·97	0·77	3	6·4	B
Oxalic	0·997*	2·00	0·60	5	8·4	A
	0·984	2·03	0·95	3	8·6	B
Malonic	0·999	2·00	1·50	5	6·5	A
	1·000	1·99	1·08	4	5·6	B
Succinic	0·996	2·06	0·62	5	6·3	A
	0·987	2·04	0·57	3	3·8	B
Glutaric	1·000	—	0·72	5	13·4	A
	0·994	2·00	0·86	3	5·8	B
Adipic	1·003	2·06	0·90	3	7·4	A
	1·004	—	0·09	3	11·7	B
Maleic	0·991	2·00	0·58	5	8·8	A
	0·996	1·95	1·20	3	7·6	B
Fumaric	0·989*	—	1·45	3	—	A
	0·994	2·03	1·20	3	7·0	B
Tartaric	0·987	1·95	0·42	3	6·6	A
	0·965	2·20	2·10	3	4·2	B

* Precipitate formed, which in the case of fumaric acid interfered.
A: Titrant, sodium methoxide; solvent, methanol for titrant and sample
B: Titrant, tetra-n-butyl ammonium hydroxide in methanol; sample in 1 : 4 methanol-benzene

rise measured with a 20–50|°C thermometer graduated in $0 \cdot 1$ degC. A precision of $\pm 0 \cdot 09\%$ was found over the range $0 \cdot 8$–$5 \cdot 5\%$ of acetic acid, compared with $\pm 0 \cdot 07\%$ with a visual titration procedure with the same reactant based on the colour change of Methyl Red at the end-point. The enthalpimetric method was stated to be superior because it was applicable to coloured samples (McClure *et al.*, 1955). A similar method from the same laboratories was developed, using the first reported continuous-flow analyser, and when this was applied to the analysis of acetic anhydride plant streams containing 1% of acetic acid the results were correct within $\pm 1\%$ (Lewis, 1957). Another continuous-flow analyser was used to determine up to 32% acetic acid in organic solvents and the results were repeatable to within $\pm 0.5\%$ of acid (Snyder, 1968).

Quilty (1967) determined the acidity in a number of different types of lubricating oils. From 0·5 to 1·0 g of the sample was dissolved in 20 ml of

Table 4.12 Titration of carboxylic acids in acetonitrile solution
(From Forman and Hume (1964))

Acid	*Apparent purity* %	*No. of* *tests*	*Precision* $\pm\%$	$-\Delta H^\circ$ kcal/mole
Benzoic	102·3	3	1·2	12·4
p-Chlorobenzoic	102·3	3	2·4	12·8
m-Chlorobenzoic	99·2	3	1·6	12·8
m-Bromobenzoic	101·3	3	1·4	12·8
m-Nitrobenzoic	99·1	3	1·1	15·4
p-Nitrobenzoic	101·4	3	2·2	13·4
Chloroacetic	101·3	2	1·6	13·6
Dichloroacetic	99·4	3	0·6	15·7
Trichloroacetic	99·1	3	2·1	19·2

propan-2-ol and titrated with 0·2 M NaOH in propan-2-ol, using a Titra-Thermo-Mat. The acidity results were compared with those obtained by a potentiometric method and to which they bore no relation, as was stated to be common when methods are compared in which these oils are used. Heats of neutralization were also determined and found to range from -1 to -28 kcal/mole. Standard solutions of phenol and benzoic acid containing from 6 to 121 mmole/kg in acid-free oils were determined to within 5 mmole/kg. It has been reported that weak acids in benzene solutions can be titrated with gaseous ammonia (Hume and Duffield, 1964).

(*ii*) PHENOLS

Parsons (1957) reported that the direct determination of phenol and tar-acids is possible with pyridine as the solvent and alcoholic potassium hydroxide as the titrant. The method was attractive enough for it to be used for routine control. Using the early titration method, Pâris and Vial (1952a) showed that if mixtures of the lower-boiling phenols were brominated, not only was their acidity increased but the increase was different for different phenols so that mixtures of certain phenols could be titrated separately. The phenols were brominated with bromate–bromide in acid solution, the resultant mixture was neutralized and the precipitated bromophenols were filtered off and dissolved in alcohol. The titrant used was 1·762 M alcoholic potassium hydroxide, and it was shown that if mixtures of phenol and the cresols were reacted and titrated, the sum of phenol + m-cresol and that of o-cresol + p-cresol could be obtained in a single titration.

(iii) VERY WEAK ACIDS

An unusual titrant has been investigated by Belisle (1971) and shown to be effective in the titration of very weak acids. Thallium(I) ethoxide, a recently available compound, is soluble in benzene and the solution is basic, and when used as a 1 M solution it gives excellent titration curves with benzoic acid and phenol; even trifluoroethanol ($pK = 12{\cdot}4$) could be titrated, but the end-point was rounded. In aprotic solvents the titration curves showed an inflection caused by the formation of an acid–anion complex. This was eliminated by the addition of small amounts of alcohol which had also to be added when the substance being titrated gave a precipitate, e.g., in the case of m-methoxybenzoic acid, in order to obtain linear titration graphs.

Two weakly acidic sulphur compounds, 2-mercapto-5-anilino-1,3,4-thiodiazol and hydrazine-N,N'-bisthiocarbonic acid allyl monoamide, have been titrated in aqueous alcoholic solution. A discontinuous manual technique was used with 1 M NaOH in 50% ethanol as titrant and the rectangular titration graph was transformed into oblique co-ordinates to allow for the influence of secondary thermal processes in the case of the monoamide. The heat of neutralization of the thiodiazol was found to be —13·80 kcal/mole and that of the two ionization stages of the mono-amide to be —5·54 and —1·64 kcal/mole respectively (Popper et $al.$ 1964).

4.1.3 *Acids in non-aqueous solution—catalytic end-point methods*

In the course of an investigation into solvents that could be used for the titration of phenols in non-aqueous solution, with a strong base, Vaughan and Swithenbank (1965) discovered that when acetone was used as the solvent any acidic material present in the acetone could be titrated without appreciable heat change but when the base was in excess it acted cataly-tically on the acetone to produce diacetone alcohol with the evolution of considerable heat. No other solvent, including ketones, was found to behave in this way. Acetaldehyde when titrated with a strong base also evolved heat but this took place immediately the base was added, irre-spective of the presence of acid. The reason for this unique behaviour of acetone is concerned with the kinetics of the three-stage reaction, in which the hydroxyl ions have a catalytic effect.

(1) CH_3—CO—CH_3 + OH^- $\rightleftharpoons$ CH_3—CO—CH_2^- + H_2O

(2) CH_3—CO—CH_2^- + CH_3—CO—CH_3
$$\rightleftharpoons CH_3\text{—CO—}CH_2\text{—}C(O^-)(CH_3)_2$$

(3) CH_3—CO—CH_2—$C(O^-)(CH_3)_2$ + H_2O
$$\rightleftharpoons CH_3COCH_2C(OH)(CH_3)_2 + OH^-$$

Sykes (1963) stated that stage (2) is slow and thus, if any acidity is present, on addition of the base titrant the neutralization of this acidity will take place more rapidly than that of stage (2) so the heat-producing reaction stage (3) does not take place until the neutralization reaction is complete. With acetaldehyde the corresponding reaction to form aldol also has three stages but the second is fast and consequently the condensation reaction takes place immediately on addition of the base.

The high enthalpy of the reaction of acetone to yield diacetone alcohol gave an exceptional 5–6 degC rise in temperature when a non-aqueous base was added to 5 ml of acetone, as can be seen in the titration of benzoic acid shown in Fig. 4.2 (curve a). For this reason the authors found that they only needed a simple titration vessel as shown in Fig. 2.16. Water was found to interfere greatly with the method as it reduced the rate of temperature rise as shown in Fig. 4.2 (curves b and c), and in order to obtain a satisfactory end-point a maximum water content of 0·2% was recommended, either in the acetone or introduced with the sample.

The effects of other solvents on the reaction were investigated and it was

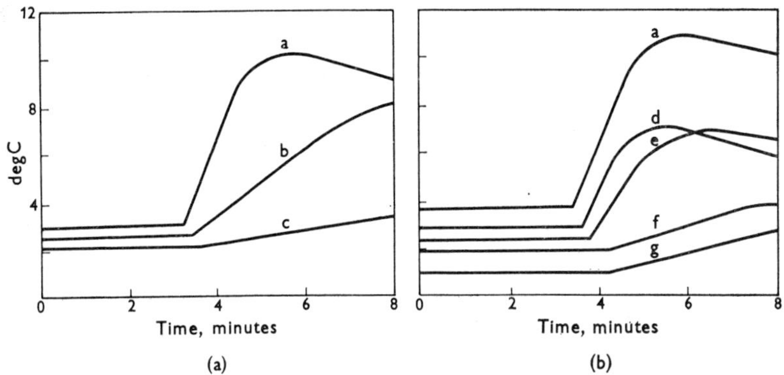

FIG. 4.2 Titration curves for benzoic acid in acetone solution.

(a) Effect of added water: curve a, dry acetone; curve b, acetone containing 0·2% water; curve c, acetone containing 2·0% water.

(b) Effect of adding solvents: curve a, dry acetone; curve d, 50% pyridine; curve e, 25% benzene; curve f, 50% benzene; curve g, 50% nitrobenzene; from Vaughan and Swithenbank (1965).

shown that benzene or nitrobenzene reduced the reaction rate but satisfactory end-points were obtained with mixtures containing not more than 25% of either (see Fig. 4.2, curves e, f, g). Pyridine did not affect the reaction rate and only reduced the overall temperature rise and could therefore be used as a pre-solvent (see Fig. 4.2, curve d). A number of base titrants were investigated and it was found that 1 M KOH in methanol was as effective as 1 M butyltrimethylammonium hydroxide in methanol, but 1 M KOH in propan-2-ol was found to give a slightly greater rate of heat evolution and was recommended for the titration.

Table 4.13 gives the wide range of acidic substances that can be titrated by the method, e.g., phenol (see curve b, Fig. 4.3), but because of the overall nature of the titration no distinction can be made between the different ionization stages of polyprotic acids. All titratable groups of the polyprotic acids were titrated except in the case of succinic acid, catechol and pyrogallol, when only $\frac{1}{2}$, $\frac{1}{2}$, and $\frac{2}{3}$ of the acidic groups, respectively, were titrated. This is most probably due to the formation of a hydrogen bond between the unionized and ionized groups of two molecules, with the form [—COO···H···OOC—]⁻. In the titration of aromatic nitrocompounds, mononitro compounds were not titratable whereas in the case of

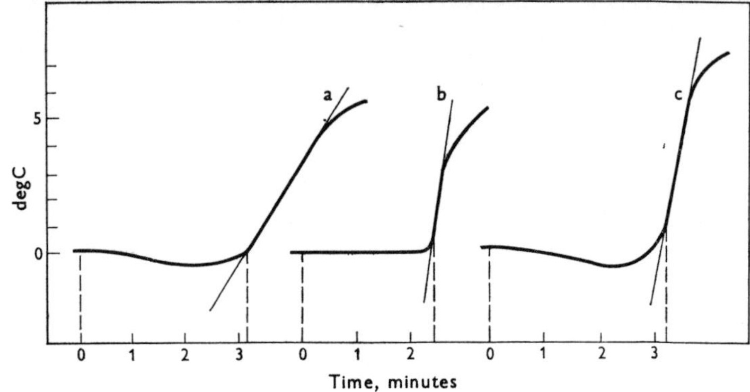

FIG. 4.3 Titration curves for: a coal; b *p*-t-butylphenol; c 2, 6-di-t-butyl-4-methylphenol. Titrant: M KOH in propan-2-ol at 0·1 ml/min; from Vaughan and Swithenbank (1970).

polynitro compounds all but one nitro group could be titrated, i.e., dinitrobenzene was found to be 'monoprotic', 1,3,5-trinitrobenzene 'diprotic' and picric acid 'triprotic'.

Table 4.13 Materials titratable by the acetone indicator method
(From Vaughan and Swithenbank (1965))

Group	*Individual acids*
Carboxylic acids (monoprotic)	acetic; stearic; benzoic
Carboxylic acids (diprotic)	oxalic; succinic; adipic; phthalic
Hydroxy-acids	tartaric; citric
Phenols (monohydric)	phenol; 2,6-xylenol; 2,6-di-t-butyl-4-methyl phenol; 2-naphthol
Phenols (polyhydric)	quinol; resorcinol; catechol; pyrogallol
Phenolic acids	salicylic; 4-hydroxybenzoic; 2-hydroxynaphthoic
Keto-enols	acetylacetone; acetoacetic ester; dimedone; diethyl malonate
Imides	succinimide; phthalimide
Aromatic nitrocompounds	*m*-dinitrobenzene; 2,5-dinitrofluorene; 1,3,5-trinitrobenzene; picric acid

The method was found to be applicable to a wide variety of substances including gums, resins, polymers, bitumens and tars, and the results of the determination of the acidity in these substances were precise to within ±1·4%.

A number of phenols containing tar-oils were examined, and the results when compared with a sodium hydroxide extraction method (S.T.P.T.C., 1967) showed good agreement at high values, but at low values higher results were obtained with the thermometric method because of difficulties with the extraction procedure.

In the titration of certain very weak acids it was found that the end-point region was very curved and under these circumstances the stoichiometric end-point occurs at the point where the tangent to the main temperature-rise leaves the titration curve at the lower temperature end. This is shown in Fig. 4.3 (curve c) for the titration of 2,6-di-t-butyl-4-methylphenol where the end-point is shown as a vertical dotted line. This curvature was attributed to the delay in the exothermic formation of the potassium salt because in the case of p-t-butylphenol the titration curve begins endothermally and recovers abruptly with the precipitation of the salt to give a level curve with a normal sharp end-point (see Fig. 2.33 (b)).

The curvature at the end-point was also found in the titration of the phenolic hydroxyl content of bituminous coals (Vaughan and Swithenbank, 1970). In this method about 30 mg of the coal were ground, first dry for five minutes and then in the presence of 0·5 ml of pyridine for one hour, in an agate ball-mill. The pyridine extract was diluted with acetone and titrated with 1 M KOH in propan-2-ol. A typical titration curve for coal is shown in Fig. 4.3 (curve a) and the curvature in the region of the end-point was attributed to the presence of highly hindered phenols in the coal, which, in the case of high carbon content coals, were not titratable by potentiometric or acetylation methods as can be seen from Table 4.14. Eight bituminous coals were examined containing between 79 and 94% of carbon and the results indicated that the phenolic hydroxyl accounts for 37–82% of the organic oxygen, depending on the rank of the coal.

Table 4.14 Comparison of thermometric and acetylation methods for phenols in bituminous coals
(From Vaughan and Swithenbank (1970))

Source	Coal rank Code no.	Carbon %	Oxygen %	Phenolic hydroxyl, %	
				Acetylation	Thermometric
Oakdale	301a	90·7	2·0	0·51	1·00
Annesley	702	83·8	7·8	4·86	4·90
Linby	802	83·1	9·2	5·65	5·78

In a paper given at a British Wood Preserving Association meeting, Vaughan (1967) reported that he had used the method to determine the phenols content of wood preserved with creosote. Small sections of the wood were extracted with acetone and the extract titrated. The results showed that the phenols had the same distribution pattern in the wood as the hydrocarbons, e.g., naphthalene, which were determined by gas-liquid chromatography.

Because the temperature rise at the end-point in the acetone indicator method is so large, a short-range thermometer can be used to detect the temperature rise and this, coupled with a burette-tipped syringe with a needle (see Section 2.1.2) and simple titration vessel, has been specified in a standard method for the titration of phenols in tar-oils (S.T.P.C.T., 1967). A similar apparatus is used in a British Standard method for the determination of tar-oils in preserved textiles (BS 2087: 1971). In this method the textile is extracted with acetone. The extract is evaporated to low volume to remove water, diluted with acetone and then titrated.

The acetone indicator method has also been used by Vajgand and his co-workers (1970, 1971) in conjunction with coulometrically generated titrant. The supporting electrolyte in this case was 0.5 M $NaClO_4$ in dry acetone and in six successive determinations of 0.8-mg samples of benzoic acid the precision obtained was $\pm 1.5\%$ in close agreement with the value of $\pm 1.4\%$ obtained by the titration method. Vaughan and Swithenbank (1965) showed that endothermic heats of reaction were obtained when a non-aqueous base was added to diacetone alcohol and Vajgand also found that a mixture of acetone with diacetone alcohol was an effective solvent for the titration.

This catalytic end-point method for the determination of acids has been considerably advanced by Greenhow and Spencer (1972a, c) who used polymerizable monomers as the solvents. These were acrylonitrile, methyl acrylate and dimethyl itaconate, and the base titrants, that acted as catalyst, included tetra-n-butylammonium hydroxide in toluene/methanol and potassium hydroxide in propan-2-ol. The heat evolved at the end-point is so large that the method is applicable to the determination of acids at the 100 ppm level, 0.001 M titrant being used.

4.1.4 *Bases in aqueous solution*

(*i*) BASIC NITROGEN COMPOUNDS

In a paper to the American Chemical Society, Parsons (1957) described the thermometric titration of pyridine in the presence of ammonia and used the method for routine control. Vaughan and Swithenbank (1967) showed that when a wide variety of basic organic nitrogen compounds were titrated with 5 M HCl they fell generally into two categories: (*a*) those with a steep titration graph indistinguishable from that of ammonia and (*b*) those with a less-steep titration graph indistinguishable from that of pyridine. Representatives of the two classes are shown in Table 4.15 and it can be seen that derivatives of ammonia fall into the first class and those of pyridine or of aniline into the second.

The reason for this division is the wide difference in pK values between the two classes and the small differences between the members within the two classes. For example the pK values for methylamine and diethylamine are 10.6 and 11.0 respectively whereas those of pyridine and aniline are 5.2 and 4.6 respectively.

Using the early titration method, Daftary and Haldar (1961) also showed that individual components of binary mixtures of ethylenediamine with pyridine or aniline, or of diethylamine with pyridine or aniline, could be titrated with acid. When 40 ml of a solution containing about 60–70mg of each base were titrated with 0·168 M HCl the results for any binary combination were correct within $\pm 0.7\%$. Vaughan and Swithenbank (1967) using 5 M HCl, determined the ammonia and pyridine base content

Table 4.15 Basic nitrogen organic compounds—distinction into two classes
(From Vaughan and Swithenbank (1967))

Steep-graph bases— titrated first	Ammonia, methylamine, diethylamine, triethylamine, pyrollidine, cyclohexylamine, ethylenediamine, ethanolamine
Less steep-graph bases— titrated second	Pyridine, 3-picoline, quinoline, aniline, *o*-toluidine

of tar products, with the results given in Table 4.16, which show good duplication in an analysis that is extremely difficult to carry out by other means, for example, by distillation from a fixed pH solution followed by acid titration (Ashmore, 1949). They found that if the samples contained a high proportion of ammonia the large temperature rise caused by the neutralization of the ammonia obscured the pyridine end-point, and in these cases they titrated the bases as weak-base salts as described in Section 4.1.1(*ix*). Daftary and Haldar (1961) analysed mixtures of bases by titration, first with acid and then with sodium nitrite solution. Details of this method are given in Section 4.2.

Table 4.16 Ammonia and pyridine base content of tar products
(From Vaughan and Swithenbank (1967))

Source	*Basic nitrogen compounds* $\%w/w$	
	Ammonia as NH_3	*Pyridine bases as* C_6H_5N
Distillate from phenols plant rectifier	2·42, 2·46	0·29, 0·33
Ammonia sulphate liquor from pyridine plant	0·78, 0·81	0·11, 0·12
Aqueous condensate from phenols plant	0·777, 0·080	0·25, 0·26
Aqueous liquor from crude tar	0·027, 0·011	nil, nil
Ammonia liquor from crude tar	0·011, 0·011	nil, nil

The precision and results obtained in the titration of individual bases by various workers are given in Table 4.17 and as will be seen the error of the methods is about $\pm 1\%$ which is very satisfactory.

The enthalpy of protonation of a large number of organic bases in aqueous solution has been determined by thermometric titrimetry and the

Table 4.17 Precision of methods for the determination of individual organic bases in aqueous solutions

Compound	Taken	Volume ml	Titrant M	No. of tests	Error %	Precision ±%	Ref.
Methylamine	17–51 mg	40	0·168	3	±0·9	—	(a)
Diethylamine	17–80 mg	40	0·168	3	±1·2	—	(a)
Ethylenediamine	11–94 mg	40	0·168	3	±0·9[2]	—	(a)
	0·005 M	100	2·00	5	−1·1[2]	0·25	(b)
	0·5024 mmole	75	0·97	5	±1·1[1]	1·9[1]	(c)
					+0·1[2]	2·0[2]	(c)
Nicotine	81–243 mg	40	0·168	3	±1·0	—	(a)
Ethanolamine	0·5024 mmole	75	0·97	4	−0·3	0·9	(c)
2-Amino-1-butanol	0·5024 mmole	75	0·97	9	−0·6	0·9	(c)
2-Methyl-2-amino-1-propanol	0·5024 mmole	75	0·97	8	−0·4	0·6	(c)
2-Methyl-2-amino-1,3-propanediol	0·5024 mmole	75	0·97	6	+0·3	0·5	(c)
Tris(hydroxymethyl)−	0·5024 mmole	75	0·97	6	+0·1	0·4	(c)
aminomethane				7	−0·2	1·2	(c)
Niacinamide	0·01 M	100	2·0	6	−1·4	0·37	(b)
Pyridine	1·186 mmole	60	0·115	7	−0·96	4·7	(d)
	40–158 mg	40	0·168	—	+0·6	1·3	(a)
	0·05–0·50*	—	0·6	5	+1·0	—	(e)
	0·39–0·99 mmole	75	0·97	5	−0·8	0·2	(c)
Aniline	46–186 mg	40	0·168	3	±0·9	—	(a)
o-Toluidine	53–111 mg	40	0·68	2	±1·0	—	(a)
m-Aminophenol	61–245 mg	40	0·68	2	±0·7	—	(a)

(a) Burette addition–Beckman thermometer (Daftary and Haldar, 1961)
(b) Syringe–thermistor (De Leo and Stern, 1966)
(c) Differential titrator (Tyson et al., 1961)
(d) Continuous-flow burette–thermistor (Linde et al., 1953)
(e) Continuous-flow titrator (Priestley et al., 1965, 1968)
[1,2] First and second ionization respectively
* Arbitrary concentration units.

results are summarized in Table 4.18. Studies of the lower aliphatic amines and hydroxylamines showed that ethanolamine is more basic than ethylamine, which is thought to be due to the effect of the hydroxyl group. The basicity of diethylamine is greater than both that of ethylamine and triethylamine, this being caused by steric effects; the enthalpy of protonation of the second amino group in ethylenediamine is lower than that of

Table 4.18 Enthalpy values for the protonation of bases in aqueous solution at 25 °C

Substance	$-\Delta H_1$ kcal/mole^{-1}	$-\Delta H_2$ kcal/mole^{-1}	
Ethylamine	11·37	—	(a)
Ethanolamine	8·20	—	(a)
Diethylamine	11·67	—	(a)
Diethanolamine	10·73	—	(a)
Triethylamine	8·30	—	(a)
Triethanolamine	7·89	—	(a)
Ethylenediamine	12·11	11·64	(a)
	12·38 ± 0·18	10·83 ± 0·11	(b)
	12·00 ± 0·18	10·89 ± 0·16	(c)
1, 3-Diaminopropane	13·19 ± 0·19	12·97 ± 0·16	(c)
Tris(hydroxymethyl)aminomethane	11·89 ± 0·10	—	(b)
	11·33 ± 0·09	—	(d)
	11·38 ± 0·06	—	(e)
	11·39 ± 0·04	—	(f)
2-Amino-1-butanol	12·82 ± 0·10	—	(b)
2-Methyl-2-amino-1-propanol	13·38 ± 0·10	—	(b)
2-Methyl-2-amino-1, 3-propanediol	12·43 ± 0·10	—	(b)
Pyridine	4·67 ± 0·07	—	
	4·98 ± 0·04	—	(f)
2-Aminomethylpyridine	10·57 ± 0·16	6·31 ± 0·09	(c)
2,2′-Aminoethylpyridine	11·60 ± 0·17	6·27 ± 0·09	(c)
Aniline	7·92	—	(g)
o-Phenylenediamine	8·77	3·87	(g)
m-Phenylenediamine	2·16	0·16	(g)
p-Phenylenediamine	8·70	2·48	(g)
Imidazole	8·78 ± 0·03	—	(f)
	9·03 ± 0·12	—	(c)
2,2′-Di-imidazole	6·08 ± 0·25	—	(c)
4(5)-Aminomethylimidazole	9·73 ± 0·14	7·43 ± 0·11	(c)
4(5)-2′-Aminoethylimidazole	10·28 ± 0·15	9·25 ± 0·14	(c)

(a) Popper *et al.* (1967); (b) Tyson *et al.* (1961); (c) Holmes and Williams (1967b); (d) Grenthe *et al.* (1970); (e) Wilson and Smith (1969); (f) Christensen *et al.* (1968b); (g) Popper *et al.* (1965).

the first and this is attributed to the hydration of the second amine group when the first is protonated (Popper *et al.*, 1967, Tyson *et al.*, 1961). Study of pairs of a substance and its higher homologue in the case of diacidic bases ((c), in Table 4.18) shows that the higher homologue has the higher enthalpy, this probably being caused by the inductive effect of the extra methylene group affecting both centres capable of protonation (Holmes and Williams, 1967b).

The results of the determination of the enthalpy of protonation of tris(hydroxymethyl)aminomethane (THAM or Tris buffer) shows very close agreement between the results of three of the workers and supports the use of this material as a standard for thermometric titrimetry as proposed by Wilson and Smith (1969).

The thermometric titration method has been used by Christensen and his co-workers (1968b) to determine pK, $\Delta S°$ and $\Delta H°$ for the protonation of basic nitrogen compounds, and examples of their results are given in Table 4.19 and show the high accuracy they obtained. In this study acetic acid was used as the titrant.

Table 4.19 Thermometric data for the protonation of some organic nitrogen compounds
(From Christensen *et al.* (1968b))

Compound	pK	$-\Delta H°$ kcal/mole	$-\Delta S°$ cal mole^{-1} degK^{-1}
Pyridine	5·168 ± 0·018	4·98 ± 0·04	−7·0 ± 0·2
Imidazole	6·986 ± 0·015	8·78 ± 0·03	−2·5 ± 0·1
Tris(hydroxymethyl)amino-methane (THAM)	8·030 ± 0·37	11·39 ± 0·04	+1·5 ± 0·2
Aminoacetic acid*	9·59 ± 0·15	10·68 ± 0·03	−8·1 ± 0·7
Metanilic acid*	3·756 ± 0·036	4·92 ± 0·15	−0·7 ± 0·7

* Titrated as their sodium salts

The early titration method was used to determine the dissociation constants and heats of neutralization of piperazine, cinchonine and quinidine. The titration curves of all three bases had two inflections, showing the bases to be diacidic.

(*ii*) WEAK-ACID SALTS

It is possible to titrate weak organic acid salts with strong acids, for example, the free sodium hydroxide and sodium phenate contents of phenol solutions to which excess of sodium hydroxide has been added can be determined by titration with sulphuric acid (Parsons, 1957). More complex mixtures of phenol, carbonate, sulphite and aromatic sulphonic acids, when made basic with sodium hydroxide, can also be analysed by titration with sulphuric acid (Vaughan, 1970). It is obvious that this method is applicable to similar systems.

4.1.5 *Bases in non-aqueous solution*

(*i*) BASIC NITROGEN COMPOUNDS—NORMAL END-POINT METHODS

A number of organic substances have been studied as solvents for the non-aqueous titration of organic bases. Forman and Hume (1959, 1964) used the aprotic solvent acetonitrile because of its favourable properties and because Fritz (1953) used it in the potentiometric titration of weak bases. They dissolved hydrogen bromide in it to make an approximately 0·1 M solution and used this as the titrant. They were unable to prepare a titrant containing perchloric acid, and hydrogen chloride was found to be less satisfactory than hydrogen bromide because of its volatility. Even the hydrogen bromide titrant needed to be standardized every day. The curve obtained in the titration of aniline by this method is given in Fig. 2.33(a).

Twenty-seven organic bases have been examined, and the analytical results and their precision and the values of the enthalpy of protonation, with the standard deviation for each compound, are given in Table 4.20. If

Table 4.20 Titration of bases with hydrogen bromide, both in acetonitrile
(From Forman and Hume (1959, 1964))

Compound	Taken mmole/ 15 ml	No. of tests	Purity %	Precision ±%	$-\Delta H°$, kcal/mole Found	Std. dev.
1° *Amines*						
Ethylamine	0·535	3	98·9	0·93	26·4	0·9
n-Propylamine	0·445	4	98·9	0·40	25·8	1·0
Isopropylamine	0·700	2	99·0	1·64	26·9	0·4
n-Butylamine	0·540	4	99·4	1·00	25·8	0·3
s-Butylamine	0·535	2	99·3	0·00	25·8	0·5
t-Butylamine	0·449	4	99·3	0·35	25·9	0·4
Isobutylamine	0·995	3	98·3	0·70	25·5	0·7
Aniline	0·930	4	98·7	0·78	14·9	0·4
p-Toluidine	0·828	4	99·5	0·23	13·7	0·3
p-Anisidine	0·640	3	99·1	0·40	17·7	0·7
p-Chloroaniline	1·028	4	100·5	1·23	12·3	0·6
m-Chloroaniline	0·820	4	104·0	1·12	11·9	0·3
p-Bromoaniline	0·794	4	103·6	0·21	11·6	0·6
2° *Amines*						
Diethylamine	0·595	3	100·5	0·40	26·8	0·5
Di-isopropylamine	0·130	3	99·2	1·56	24·4	0·4
Di-n-butylamine	0·600	4	99·7	0·14	26·2	0·7
Di-s-butylamine	0·285	4	98·6	0·76	24·1	0·9
Morpholine	0·424	4	99·1	0·85	25·5	0·8
N-Methylaniline	0·640	3	98·1	1·14	13·2	0·7
1-Ethylpiperidine	0·111	4	99·1	0·97	24·2	0·5
Diphenylguanidine	0·708	4	99·7	0·57	20·7	0·7
3° *Amines*						
Triethylamine	0·439	4	99·3	1·10	25·2	0·4
Tri-n-butylamine	0·598	4	99·5	0·56	23·9	0·9
Tri-isobutylamine	0·365	4	98·9	0·26	18·1	0·1
N,N-Dimethylaniline	0·610	3	99·0	0·94	14·9	0·8
N,N-Diethylaniline	0·457	4	99·1	1·03	15·6	0·4
Pyridine	0·590	4	99·0	0·74	14·4	0·4

these results are compared with those given in Tables 4.17 and 4.18 it can be seen that the overall precision is poorer than that obtained in aqueous solution, but Forman and Hume pointed out that the precision was similar to that reported by Fritz for the potentiometric titrations in the same solvent. Attempts to titrate *o*-chloroaniline and *m*- and *p*-nitroanilines were unsuccessful.

The same authors also showed that the individual components of mixtures of pyridine with either n-butylamine or diphenylguanidine could be titrated separately in a single titration. The pyridine–diphenyl-guanidine mixture gave a poorer middle inflection because of the lower ratio of the heats of protonation of the two bases (14·4 kcal/mole and

Table 4.21 Titration of bases with trichloroacetic acid, both in benzene solution
(Enthalpies of reaction; from Mead (1962))

Compound	$-\Delta H_1$* kcal/mole	Std. dev. $\pm$kcal/mole	No. of tests	$-\Delta H_2$* kcal/mole
1° Amines				
n-Butylamine	29·3	0·0	3	0·7
s-Butylamine	29·6	0·6	8	0·3
Allylamine	27·4	0·2	4	0·4
Benzylamine	26·6	0·1	4	0·6
p-Anisidine	18·8	0·3	3	0·0
Aniline	16·4	0·2	3	—
m-Nitroaniline	0·3	—	1	—
m-Bromoaniline	6·2	0·2	2	—
p-Chloroaniline	9·8	0·3	3	—
2° Amines				
Di-isopropylamine	29·4	0·1	4	1·2
Di-n-butylamine	33·3	0·4	3	1·9
Di-s-butylamine	28·5	0·3	8	1·3
Di-isobutylamine	27·2	0·4	4	2·9
Diallylamine	24·5	0·2	4	1·8
Piperidine	28·3	0·3	5	2·3
Morpholine	25·0	0·5	4	2·3
Dibenzylamine	22·5	0·2	4	2·6
N-Methylaniline	13·8	0·1	4	—
1,3-Diphenylguanidine	23·8	0·2	4	1·9
Tetramethylguanidine	31·2	0·2	4	3·3
N-Ethylaniline	16·5	0·2	3	—
Di-n-propylamine	29·8	0·3	4	2·0
Isoquinoline	12·7	0·4	4	—
Bis(cyanomethyl)amine	0·8	—	1	—
3° Amines				
Tri-n-butylamine	23·0	0·5	4	4·9
Tribenzylamine	13·5	0·3	2	—
Dimethylbenzylamine	19·8	0·2	4	4·8
N,*N*-Dimethylaniline	11·3	0·5	4	—
N,*N*-Diethylaniline	15·3	0·4	3	—
Pyridine	13·6	0·4	3	—
Triethylenediamine	21·6	0·2	3	—
Tetramethylethylenediamine	22·2	0·1	3	—
Triethanolamine	27·8	0·8	2	—
p-Chloro-*N*,*N*-diethylaniline	12·7	0·6	3	—

* $-\Delta H_1$, 1 : 1 amine:trichloroacetic acid; $-\Delta H_2$, 1 : 2 amine:trichloroacetic acid.

20·7 kcal/mole) compared to the mixture of pyridine with n-butylamine (14·4 kcal/mole and 25·8 kcal/mole).

Solvation of a base molecule or its conjugate acid by water molecules interferes with the comparison of the dissociation constants of primary, secondary and tertiary amines and because of this Mead (1962) chose the non-solvating aprotic solvent, benzene, for his studies of the heats of neutralization of thirty-five amines. The amines, 0·02 M in 50 ml of solution, were titrated at 28 °C with 1·0 M trichloroacetic acid, also in benzene solution, and it was found that a number of bases not only showed 1 : 1 amine: trichloroacetic acid reactions but also 1 : 2 reactions. The results obtained are given in Table 4.21, in which the column headed $-\Delta H_1$ is the heat of the 1 : 1 reaction and $-\Delta H_2$ that of the 1 : 2 reaction. Mead showed that when the $-\Delta H_1$ results were plotted against the pK values measured in aqueous solution, the primary, secondary and tertiary amines showed three parallel lines representing each of the three classes

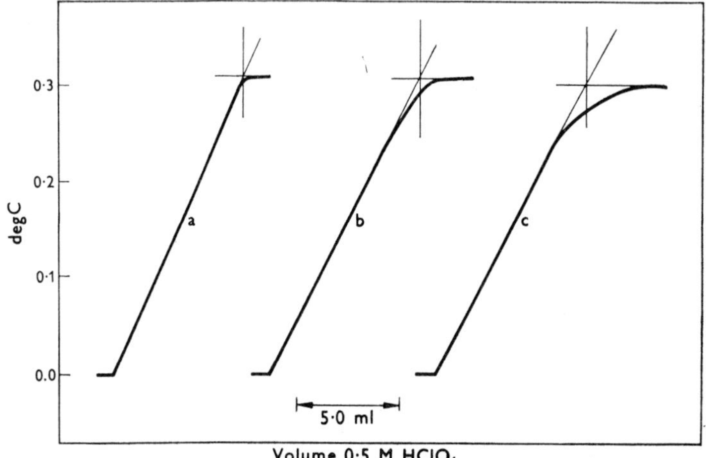

FIG. 4.4 Titration of very weak bases with 0·5 M perchloric acid, both in anhydrous acetic acid.

a urea; b acetamide; c acetanilide; from Keily and Hume (1964).

and this was attributed to the differences in solvation of the bases by benzene and water. In this solvent–titrant system *m*-nitroaniline could be titrated whereas in the system acetonitrile–hydrogen bromide it could not.

Acids are excellent solvents for the titration of bases and acetic acid as solvent with perchloric acid as titrant has been extensively used for both the indicator and potentiometric titration of a great variety of bases. It is therefore not surprising that the system has been studied as a thermometric method. Keily and Hume (1964) used perchloric acid in acetic acid which was either anhydrous, or contained sufficient water to be in 1 : 1 molar ratio with the perchloric acid present. They called the latter titrant solution 'hydronium perchlorate'. Hall (1930), using the potentiometric method, classified bases into three broad categories according to the magnitude of the potential break on titration with perchloric acid. Keily and Hume selected bases from these three categories to study their thermometric titration with the two titrants and in order to eliminate heats of mixing used a differential titration system.

Aniline, diethylaniline, n-butylamine, diphenylguanidine and pyridine, representing strong bases, were found to be titrated successfully with both titrants, the end-points being sharp. Diphenylamine and *p*-nitroaniline, classified by Hall as intermediate-strength bases, gave sharp end-points with the anhydrous titrant but with the 'hydronium perchlorate' gave rounded end-points which were amenable to extrapolation. With urea, acetamide and acetanilide, representing the very weak bases, the anhydrous titrant gave a sharp end-point with urea and rounded end-points with the other two bases, but these were again amenable to extrapolation.

Table 4.22 Determination of purity and heat of reaction of organic bases with perchloric acid in acetic acid solution

(From Keily and Hume (1964))

Base	Purity %	No. of tests	Precision ±%	Heat of reaction $-\Delta H$ kcal/mole	Std. dev. ±kcal/mole
Strong bases					
Aniline	98·4	2	0·35	5·2	0·10
	98·8[1]	2	0·14	7·8	0·50
Diethylaniline	99·7	2	0·14	5·4	0·14
	99·8[1]	2	0·28	8·0	0·00
n-Butylamine	96·4	2	0·14	4·7(7·4)[2]	0·14
Diphenylguanidine	100·0	2	0·77	4·1(6·8)[2]	0·14
Pyridine	99·3[1]	2	0·10	8·0	0·21
Intermediate bases					
Diphenylamine	99·6	2	0·49	4·0(6·7)[2]	0·14
p-Nitroaniline	98·9	2	0·28	6·4(9·1)[2]	0·28
Weak bases					
Urea	100·1	6	0·44	1·0	0·31
	99·7[1]	3	0·25	3·3	0·12
Acetamide	99·9[1]	3	0·81	3·9	0·14
Acetanilide	103[1]	3	3·9	3·9	0·14

[1] Anhydrous titrant
[2] Values in anhydrous system predicted by adding 4·7 kcal/mole

Titration with 'hydronium perchlorate' gave poor end-points. The titration curves of these three weak bases, which cannot be titrated by the potentiometric method, are shown in Fig. 4.4.

Keily and Hume also determined the heat of reaction of perchloric acid with the bases (approximately 0·05 M in 50 ml of solution), using the initial slope method, and these results together with the analytical results are given in Table 4.22.

It is interesting to compare the results obtained in the determination of the heats of neutralization of the same base in different solvents. For example, those of aniline are -5.2, -7.9, -14.9 and -16.4 kcal/mole in acetic acid, water, acetonitrile and benzene respectively. This shows that the reaction, $H \cdot Solvent^+ + H_2O \rightarrow H_3O^+ + Solvent$, must be exothermic.

Several aliphatic amines present in air to the extent of 0.1–1.0% by volume have been determined by titration with gaseous hydrogen chloride (Hume and Duffield, 1964).

(*ii*) BASIC NITROGEN COMPOUNDS—CATALYSED END-POINT METHODS

A new and valuable method for the determination of bases in acetic acid solution has been developed by Vajgand and his co-workers (Vajgand and Gaál, 1966, 1967; Vajgand *et al.*, 1966, 1968) as a result of a suggestion in a paper by Keily and Hume (1964). The method is based on the addition of approximately 2% of water and 8% of acetic anhydride to the acetic acid solvent. The titrant used was 0.1 M perchloric acid in acetic acid and under these conditions, when any bases present had been titrated, the

Table 4.23 Determination of bases in acetic acid solution, by the catalysed end-point method

(From Vajgand *et al.* (1966; 1967; 1968))

Compound	Taken mg	Found mg	No. of tests	Precision ±%
Aminopyrine	0·1500	0·1496	6	0·29
Antipyrine	0·1000	0·1046	6	0·53
	0·2000	0·1942	8	0·11
Brucine	0·1000	0·1214	5	0·43
	0·1500	0·1518	6	0·18
Dimethylaniline	0·0500	0·0490	3	0·26
Cinchonine	0·1000	0·0970	9	0·21
Flagyle[1]	0·2000	0·1985	5	0·42
Inversal[2]	0·2000	0·1858	10	0·08
Triethylamine	0·1000	0·0965	3	0·36
	0·1500	0·1503	6	0·07

[1] β-Oxyethyl-2-methyl-5-nitroimidazole
[2] *p*-Benzoquinone–amidinohydrazonethiosemicarbazone

slight excess of perchloric acid catalysed the violent reaction of water with the acetic anhydride which had been added and which reacted only slowly in the absence of perchloric acid. The titrant was added incrementally at the rate of 0.1 ml/min and the volume added was plotted against the logarithm of the temperature rise; the portion of the titration curve after the end-point was linear. The results obtained are given in Table 4.32 and show that the method is reasonably accurate and precise. In the later paper

(1968) the authors also reported a continuous titration procedure, which gave similar results. Both procedures gave results which compared favourably with those obtained by potentiometric titration with perchloric acid.

Vajgand and his co-workers (1968; 1970; 1971) also used a coulometrically generated titrant in a similar type of titration. The supporting electrolyte consisted of 0.1 M $NaClO_4$ in a $1:7$ mixture of acetic acid and acetic anhydride that contained 0.2% of water. They investigated the titration of triethylamine, dimethylaniline, pyridine, 4-picoline, quinoline, antipyrine and caffeine, and for 1.30–3.30 mg of sample, results were obtained with a mean deviation of 0.8% or less, even with the exceptionally weak base caffeine ($pK = 13.39$). Similar results were obtained when acetic anhydride was replaced by propionic anhydride. In the last paper they described another catalysed end-point method based on the acetylation of hydroquinone with acetic anhydride, again catalysed with perchloric acid, in which acetic or propionic acid was used as the solvent. With this system the solvent can be completely anhydrous and consequently extremely weak bases can be titrated. The electrolyte used in this case was 0.1 M $NaClO_4$ in a solvent which was either acetic acid + acetic anhydride ($1:7$), nitrobenzene + acetic anhydride ($4:1$), formic acid + acetic anhydride ($1:9$) or acetic anhydride. Good agreement was obtained in the determination of pyridine and other bases by using the same apparatus and detecting the end-point either thermometrically, potentiometrically or photometrically.

A similar anhydrous titration system has also been investigated by Goizman (1969). The base was dissolved in acetic anhydride, the bulk of the 0.1 M $HClO_4$ (in acetic acid) titrant was added, followed by the reacting hydroxyl compound, and the titration was completed. The reactants studied were methanol, ethanol, propanol, isopropyl alcohol, butanol, isobutyl alcohol, pentanol, octanol, glycol and phenol; the bases titrated were potassium hydrogen phthalate, isonicotinyl (3-methoxy-4-hydroxybenzylidene) hydrazine and 2-ethoxy-6,9-diaminoacridine lactate (solasodine).

This catalytic end-point method for the determination of bases has been considerably advanced by Greenhow and Spencer (1972b, c) who used polymerizable monomers as the solvent. These were methyl styrene and isobutyl vinyl ether, and the acid titrants, that acted as catalysts, were perchloric acid in acetic acid and, more usually, boron trifluoride etherate in dioxan. The heat evolved at the end-point is so great that the method is applicable to the determination of bases at the 10 ppm level with 0.001 M titrant. The method was found to be applicable to all types of amines, amides and some basic sulphur and phosphorus compounds.

(*iii*) BASIC NITROGEN COMPOUNDS—HEAT OF DILUTION END-POINT METHODS

In an investigation into the possibility of utilizing the heat of dilution for end-point enhancement Vaughan and Swithenbank (1967) found that when hydrogen chloride was dissolved in alcohols, the solutions gave endothermic heats of dilution when titrated into a wide variety of solvents. Table 4.24 gives the initial temperature change when 5 M hydrogen chloride in propan-2-ol was added at a rate of 0·0011 ml/s to 5 ml of a selection of solvents, including water.

Figure 4.5 shows the titration graphs obtained when 5 ml of (*a*) acetone and (*b*) benzene were titrated with the titrant and (*c*) and (*d*) the same solvents containing 30 mg of pyridine. The graphs show that the base is

Table 4.24 Initial temperature change on titration of 5 M HCl in propan-2-ol into 5 ml of solvent
(From Vaughan and Swithenbank (1967))

Solvent	Initial temperature change, degC/min	Solvent	Initial temperature change, degC/min
Carbon tetrachloride	−2·30	Diethyl ether	−0·72
Benzene	−1·56	Acetic acid	−0·70
Nitrobenzene	−1·56	Petroleum spirit (60–120°)	−0·60
Acetone	−0·90	Methanol	+0·12
Dioxan	−0·82	Propan-2-ol	+0·16
Isobutyl methyl ketone	−0·72	Water	+0·65

titrated first with a temperature rise, caused by the neutralization of the base, followed by a rapid temperature fall at the end-point, caused by the endothermic heat of dilution of the titrant. From this work it became apparent that a wide range of solvents, except alcohols, could be used for the thermometric titration of bases and that water should be absent from

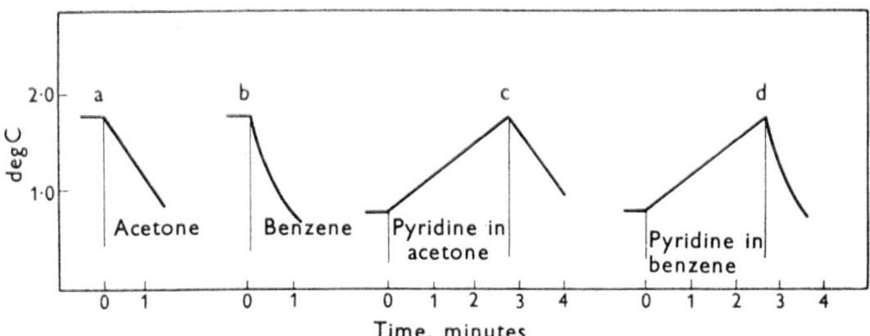

FIG. 4.5 Graphs of thermometric titrations in non-aqueous solvents with 5 M HCl in propan-2-ol; from Vaughan and Swithenbank (1967).

those solvents with which it is miscible. Thirty-six bases of all types, as given in Table 4.25, were successfully titrated in all the solvents, but a number of weak bases were only titratable in acetic acid solution as shown by (*a*) in Table 4.25 and Vaughan and Swithenbank used this to distinguish different types of bases in admixture. For example if a mixture of aniline and *p*-nitroaniline were titrated in acetone solution only the aniline could be titrated, whereas in acetic acid solution both could be titrated, the titration graph showing an inflection indicating that the aniline was titrated first, followed by the *p*-nitroaniline with a more exothermic reaction.

Table 4.25 Bases titratable with hydrogen chloride in propan-2-ol, using a heat of dilution end-point enhancement method
(From Vaughan and Swithenbank (1967))

ALIPHATIC AMINES

Ethylamine; n-octylamine; monoethanolamine; brucine; diethylamine; piperidine; trimethylamine; nylon

AROMATIC AMINES

Primary: anilinc; *o*-toluidine; 1-naphthylamine; 2-aminoanthracene; *o*-phenetidine; *o*-aminophenol; *p*-aminobenzoic acid; *m*-nitroaniline (*a*); *p*-nitroaniline (*a*); *p*-phenylenediamine; benzidine
Secondary: *N*-ethylaniline; diphenylamine (*a*); diphenylbenzidine (*a*); benzotriazole (*a*).
Tertiary: *N*, *N'*-diethylaniline; pyridine; 2-picoline; quinoline; acridine; 2,2'-bipyridyl; 2-hydroxpyridine; 8-hydroxyquinoline; quinaldinic acid; 2,2'-pyridylamine; 2-(2-aminoethyl)pyridine

AMIDES

Urea; thiourea (*a*); 6-hydroxycaproylhydrazide

(*a*) Titratable only in acetic acid solution

The method was used to determine the basic nitrogen content of a variety of tar products and the results obtained are compared in Table 4.26 with those from a potentiometric titration method in which perchloric acid in acetic acid solution was used as the titrant. Because fluid tar products could act as their own 'indicator' in the titration the authors showed that it was possible to titrate the basic nitrogen in these materials directly. In these determinations about 0·5 g of the sample was dissolved in the solvent or 5 ml of the sample was taken if it were fluid. It was also found that when nylon was dissolved in *m*-cresol and a small quantity of acetone added to reduce the viscosity of the solution then the $-NH_2$ groups on the end of the nylon chains could be titrated (Vaughan, 1970).

(*iv*) WEAK-ACID SALTS

In acidic solvents weak-acid salts behave as weak Brønsted bases and can be titrated with acids. Keily and Hume (1964) used acetic acid as the solvent and perchloric acid in acetic acid as the titrant, the method being similar to that described in Section 4.1.4(*i*). They found that mercury(II) acetate was virtually neutral in acetic acid solution and because of this, if it were added in slight excess to any chloride soluble in acetic acid, e.g., lithium chloride, then insoluble mercury(II) chloride was formed and the CH_3COO^- ion thus 'metathesized' could be titrated. In the case of sodium chloride,

Table 4.26 Basic nitrogen contents of tar products, determined in several solvents

(Comparison of results; from Vaughan and Swithenbank (1967))

Product	Solvent	Basic nitrogen, % w/w	
		Thermometric	*Potentiometric*
Creosote (i)	Acetone	0·31, 0·35	0·35
Creosote (ii)	Acetone	0.79, 0·79	0·79
Anthracene oil	Acetone	0·39, 0·41	0·40
Road tar (i)	Nitrobenzene	0·38, 0·42	0·42
Road tar (ii)	Nitrobenzene	0·57, 0·58	0·61
Phenol distillation residue	Acetic acid	0·44, 0·44	0·47
High-boiling tar acids (A)	Acetic acid	0·190, 0·186	0·180
High-boiling tar acids (A), purified	Acetic acid	0·019, 0·019	0·019
Cresylic acid	Acetone	0·014, 0·015	0·013
Benzene extract of cresylate	None	1·25, 1·29	1·22
Carbolic oil	None	0·60, 0·60	0·60
Crude naphtha	None	0·21, 0·23	0·23
Refined naphtha	None	0·071, 0·072	0·068
Xylenols	None*	0·019, 0·019	0·019

* 1 ml of acetone added to reduce the viscosity.

however, this indirect method can only be used with the 'hydronium perchlorate' titrant. Vajgand and his co-workers also used this solvent–titrant system but employed the catalysed end-point method as described above. The results obtained in the titration of acetates and other organic acid salts are given in Table 4.27, shown overleaf.

4.2 Amines, amides and anilides

The titration of the compounds described in this section is concerned with their properties other than their basicity, and titrations involving the latter are described in Sections 4.1.2 and 4.1.3. The extensive study of the reaction of copper with amines shows that copper(II) salts are possible titrants for the determination of these compounds (see Section 3.30.7).

In a paper given to the American Chemical Society in 1957 Parsons described the titration of benzthiazole and the sodium salt of sulphanilic acid with nitrous acid. Jordan and his co-workers (1966) used this reaction and found that 2 ml of 1–100 mM sulphanilamide in 0·5 M HCl could be determined with an error and precision of $\pm 1\%$ when titrated with 0·1 M $NaNO_2$. They stated that the reaction had an enthalpy of between -18 and -20 kcal/mole and that the lowest concentration determinable within an error of $\pm 1\%$ was 1 mM. Takeuchi and Yamazaki (1969b), titrating a number of aromatic primary amines, used potassium bromide as a catalyst. They obtained an error and precision of $\pm 0\cdot5$–3% and ± 2–4% respectively in the titration of 10 mM aromatic amines and a precision of $\pm 1\cdot8\%$ in the titration of 3 mM sulphanilic acid. Daftary and Haldar (1961), using the early titration method, showed that aniline in

Table 4.27 Titration of acetates and other weak-acid salts with perchloric acid, both in acetic acid solution

Compound	Purity %	Precision $\pm\%$	No. of tests	Enthalpy of reaction $-\Delta H$ kcal/mole	Ref.
Barium acetate[1]	100·0	0·23	3	7·0 ± 0·12	(a)
Lithium acetate	93·8	0·47	3	6·2 ± 0·07	(a)
Lithium chloride[2]	100·1	—	1	6·1	(a)
Sodium acetate	99·6	0·30	3	5·7 ± 0·16	(a)
	100·1	0·23	6	—	(c)
	100·1	0·00	5	—	(b)
Potassium acetate	99·6	0·16	3	7·1 ± 0·07	(a)
	100·0	0·30	6	—	(b)
	99·6	0·33	6	—	(b)
Cadmium acetate	99·8	0·10	4	7·1 ± 0·07[3]	(a)
Magnesium acetate	101·5	0·28	2	5·0 ± 0·14[3]	(a)
Lead acetate	100·0	0·21	3	7·2 ± 0·30[3]	(a)
Sodium formate	99·0	0·17	6	—	(b)
Sodium benzoate	99·8	0·25	3	—	(a)
	101·3	0·45	6	—	(c)
	99·8	0·03	6	—	(b)
Sodium citrate	98·6	0·57	3	—	(a)
Potassium hydrogen	100·0	0·26	3	—	(a)
phthalate	100·8	0·21	3	—	(b)
Sodium salicylate	100·4	0·11	6	—	(b)
Intra-iodine[4]	100·0	0·06	5	—	(b)
Pyridoxime-HCl[5]	100·8	0·21	9	—	(b)

[1] Used as a standard
[2] 'Metathesized' with mercury(II) acetate
[3] First acetate group only
[4] 1,3-Triethylammonium-2-propanol iodide
[5] 5-Hydroxy-6-methyl-3, 4-pyridine dimethanol hydrochloride
(a) Keily and Hume (1964 and also 1956)
(b) Vajgand and Gaál (1966, 1967); Vajgand et al. (1966)
(c) Vajgand et al. (1970)

admixture with pyridine could be titrated with sodium nitrite solution after the mixture had been titrated with acid, a titration in which these bases could not be distinguished. Similar results were obtained with diethylamine or ethylenediamine in the place of the aniline, but the titration of the ethylenediamine had to be carried out rapidly, or spurious results were obtained.

In the acetylation of amines or their salts with acetic anhydride, pyridine is commonly added to prevent the interference of acids present or formed, and Somiya (1932) used this procedure in the determination of aniline or its salts, using the early titration method. The titrant used was 2·4 N acetic anhydride in an acetic acid–pyridine mixture (230 ml of acetic anhydride + 335 ml of acetic acid + 335 ml of pyridine). The results obtained in the titration of approximately 7 g of a number of aniline salts are given in Table 4.28.

Table 4.28 Determination of aniline salts by acetylation

(From Somiya (1932), using the early titration method)

Compound	Aniline	
	% found	% added
Aniline sulphate (1)	64·5	65·5
Aniline sulphate (2)	64·8	65·5
Aniline hydrochloride (1)	72·1	71·8
Aniline hydrochloride (2)	71·7	71·8
Aniline hydrochloride (3)	72·2	71·8
Aniline nitrate	59·9	59·6

Monn (1966) has proposed the use of the reaction of the amino group with carbon disulphide. The compounds formed from both mono and di-substituted amines, dithiocarbamic acids, can then be titrated with silver nitrate in propan-2-ol to within 0·6%.

De Leo and Stern (1964, 1966) showed that theophylline in admixture with ethylenediamine (aminophylline) could be titrated with silver nitrate. Sharp end-points were obtained when 5 mM theophylline was titrated with 1 M $AgNO_3$, and good results were obtained when the method was applied to tablets containing the drug. Five replicate determinations on the pure substance gave a value of 99·9% ± 0·8 compared with 99·2% ± 0·4 by an indicator titration method.

Sodium hypochlorite (0·5 M) was used as a titrant by Schäfer and Wilde (1949) in an early titration method for the determination of sulphon-amides, in which the product was the compound $R \cdot SO_3$—NClNa. The compounds investigated were benzenesulphonamide; o-, m- and p-toluene-sulphonamides; o- and p-benzoic acid sulphonamides and 1-methyl-benzene-2,4-disulphonamide. Not all the reactions were exothermic and when 1 g of sample was taken the results were found to be correct within ±1%. The reactant, sodium tetraphenylborate, used as a titrant in the determination of potassium (see Section 3.3.2) has been applied by Bark and Grime (1972) to the determination of nitrogen-containing bases, by both thermometric and enthalpimetric procedures. In these methods the known excess of tetraphenylborate added was back-titrated with standard potassium chloride solution. The method was applied to the determination of milligram quantities of the physiologically active alkaloids, atropine, quinine and morphine.

The enthalpy values for the adduct formed between N,N-dimethyl-acetamide and iodine have been determined in benzene and in methylene chloride solution (Drago *et al.*, 1966). In carbon tetrachloride solution a value of $-2\cdot2$ kcal/mole was found for the heat of hydrogen bonding with the methylene chloride used as the solvent. The enthalpy values have also been obtained for the oxidation reaction of thiourea and its symmetrically N-substituted derivatives. The oxidant was cerium(IV), and in hydrochloric acid media ($0\cdot3$–$2\cdot0$ M), thiourea and ethylenethiourea were titrated correctly to within $\pm0\cdot3\%$, and in sulphuric acid media ($0\cdot3$–$0\cdot6$ M) the results for N,N'-ethylthiourea were within $\pm0\cdot4\%$ (Alexander *et al.*, 1970). The cadmium(II) thiourea complexes were investigated by Siddhanta, and these are described in Section 3.40.4(*ii*).

4.3 Anhydrides

Richmond and Eggleston (1926) assayed acetic anhydride, using a direct-injection method. In this method, 186 ml of toluene and 12 ml of aniline were placed in a Dewar flask and 2 ml of sample were injected. Somiya (1929a) titrated acetic anhydride either alone or as an excess from an acetylating reaction, with $2\cdot74$ N aniline, both being in tetrachloroethane solution, using the early titration method. It was found that better results

**Table 4.29 Comparison of enthalpimetric and thermometric methods
for acetic anhydride**
(From Greathouse *et al.* (1956))

Taken	Found, %	
%	Enthalpimetric method	Thermometric method
8·76	8·76	8·91
12·17	12·00	12·15
13·67	13·70	13·55
16·15	16·03	16·20
18·62	18·60	18·58
21·19	21·26	21·15
23·57	23·81	23·45
26·18	26·18	26·10

were obtained if an excess of aniline was added and then titrated with $2\cdot87$ N acetic anhydride. In the presence of mineral acids, which retarded the reaction, pyridine was added to eliminate their effect by its catalytic action. If the tetrachloroethane were replaced by a mixture of toluene and petroleum ether, in which the acetanilide product was insoluble, the precipitation of the anilide increased the enthalpy of the reaction and sharper end-points were obtained. Somiya obtained assay values of $93\cdot3$, $93\cdot3$, $93\cdot5$ and $93\cdot4\%$ for a sample of acetic anhydride, the last figure being obtained with the toluene solvent.

Greathouse, Janssen and Haydel (1956) determined acetic anhydride in the presence of acetic acid by reacting it with water and using perchloric acid as a catalyst. Both direct-injection and the early titration procedures were used; in the former, 200 ml of sample were placed in a Dewar flask, 4 ml of 2 M $HClO_4$ in acetic acid were added and then 25 ml of water were injected, slowly at first to avoid too much dilution of the catalyst; in the latter, a 100-ml sample containing 2 ml of the catalyst solution was titrated with 2 M H_2O in acetic acid. Results obtained on the same samples by the two methods are compared in Table 4.29.

The thermometric method was used to control a reaction in which cellulose was acetylated (Greathouse, 1957). Somiya (1929b, c), controlling the same reaction, added aniline to the sample and titrated the excess with acetic anhydride as described above, but stated that the result included both the acetic anhydride and acetyl sulphuric acid contents.

4.4 Carbonyl compounds

Ledesma and Reynolds (1968) reported that they had determined aldehydes and ketones by the reaction of the carbonyl group with 0·4 M methylphenylhydrazine. The samples were dissolved in acetic acid–benzene mixture and titrated with the reagent in benzene solution, in a differential apparatus. The method was found to be applicable to aliphatic and many aromatic aldehydes and to aliphatic ketones. The only interference was found to be with acetals and ketals, which hydrolyse to reactive carbonyl compounds, but free acid, if present, was quantitatively titrated by the base before it reacted with the carbonyl group. The minimum determinable concentration of carbonyl was found to be 2·5 mM for an aliphatic compound and 18 mM for an aromatic aldehyde and the precision and error for both was found to be $\pm 1·5\%$. In another method Bark and Bate (1972), reacted the carbonyl compound with a known excess of 2,4-dinitrophenylhydrazine, and after stirring for 10–15 minutes titrated the excess with the stable aldehyde 2-methoxybenzaldehyde. The solvent used for both titrant and titrand was isobutyl alcohol containing 4% of both water and sulphuric acid. The total time needed for this indirect method was less than twenty minutes and aromatic aldehydes were determined with an error of $\pm 1\%$.

Dilke and Eley (1949) investigated the Gatterman-Koch reaction by injecting the aldehyde into dry chlorobenzene containing a known amount of aluminium chloride and then interpreting the temperature–time graph obtained. Aqueous solutions of acetone have been analysed by injecting the sample into sulphuric acid in a similar method to that described in Section 4.8.1 (Spink and Spink, 1968), and it has been shown that furfural and salicylaldehyde can be titrated with tetramethylenediamine and that alcohol and acetone do not interfere (Korenman *et al.*, 1961).

4.5 Carboxylic acids

The titration of carboxylic acids with bases is discussed in Sections 4.1.1 and 4.1.2, but other titrants have also been used. Jordan and Ben-Yair (1957), investigating the copper(II), zinc and cadmium complexes of citric and tartaric acid, found that the nitrate salts of these metals were excellent titrants for these acids. Approximately 200 ml of about 0·02 M acid were titrated with 0·1 M $M(NO_3)_2$ by the early titration method. The results obtained are given in Table 4.30, the two results for the zinc titration showing that 1 : 1 and 1 : 2 zinc : tartaric acid complexes were formed. If a mixture of citric and tartaric acids were titrated with copper(II) nitrate the titration volume represented the sum of the acids, but if the titrand were then titrated with sodium hydroxide one equivalent of this was equal to one mole of citrate or two moles of tartrate. It is therefore possible to analyse mixtures of the two acids in this manner and the authors showed that 10 : 1 and 1 : 0 mixtures could be analysed with a precision and error of $< \pm 3\%$ and $< \pm 4\%$ respectively (see also Ben-Yair and Jordan, 1951).

Table 4.30 Determination of citrate and tartrate by titration with metal nitrates
(From Jordan and Ben-Yair (1957))

Titrant	Citrate determination		Tartrate determination	
	Precision $\pm\%$	Error $\pm\%$	Precision $\pm\%$	Error $\pm\%$
Cadmium nitrate	0·5	3	2	2
Copper(II) nitrate	2	2	0·5	0·7
			0·5	1*
Zinc nitrate	1	2	1	0·7†

* 1st break
† 2nd break

Mayr and Fisch (1929), using the early titration method, showed that oxalic acid in sulphuric acid media could be titrated with 0·1 N KMnOJ. Difficulty was encountered because of the slowness of the reaction and to avoid this it is normally carried out at 55 °C, but when 10 drops of saturated manganese(II) sulphate were added its catalytic action enabled the normal titration rate to be employed. When a titration of 3·97 ml was expected, values of 3·88 ml and 3·90 ml were obtained. The reaction has a high enthalpy ($T = 1·29$ degC/mmole), so the insoluble nature of potassium permangate does not detract from its use as a titrant, and even 0·05 N solutions were used as titrants by these authors.

The iron and aluminium complexes of oxalic, malonic, tartaric and citric acids have been investigated, as have the thorium complex of oxalic acid and the zinc and cadmium complexes of tartaric acid. These are discussed under the appropriate cation in Chapter 3.

4.6 Esters and ethers

Aqueous solutions of methyl acetate have been analysed by injecting the sample into sulphuric acid in a manner similar to that described in

Section 4.8.1 (Spink and Spink, 1968). Papoff and Zambonin (1967) have used their quasi-adiabatic thermometric method to study the kinetics of the base hydrolysis of methyl acetate and isopropyl acetate at different ratios of ester concentration.

A number of tetrahydrofurans and pyrans have been investigated as Lewis bases reacting with the tin(IV) chloride Lewis acid, and these are described in Section 3.16.4(*ii*).

4.7 Grignard reagents

Both thermometric and enthalpimetric methods were used by Parker and Vlismas (1968) to determine Grignard reagent activity. In the former, 2 ml of the Grignard reagent were added to the titration vessel, where it reacted with the air and water vapour in the vessel, the excess then being titrated with 1 M propan-2-ol in toluene. A further 2 ml of the Grignard reagent were then added and the titration continued, and the activity calculated from the two titration volumes. In the enthalpimetric method 40 ml of 2 M propan-2-ol in diethyl ether were added to the titration vessel and 0·8 ml of the Grignard reagent was injected ($T = 14\cdot2$ degC/g phenyl magnesium bromide). It was stated that in this method it was important that the Grignard reagent should be in solution in the same solvent as the titrant, i.e., diethyl ether, to eliminate heats of mixing. Table 4.31 gives the results obtained with a number of Grignard reagents, using the thermometric titration procedure, compared with those obtained by the acidimetric double titration method.

Table 4.31 Thermometric titration of Grignard reagents
(Comparison of methods; from Parker and Vlismas (1968))

Grignard reagent tested	*Grignard reagent, %*	
	Thermometric titration	*Acidimetric titration*
Methyl magnesium bromide	23·8	23·0
Methyl magnesium iodide	10·1	9·9
Ethyl magnesium iodide	41·5	41·4
Phenyl magnesium bromide	19·1	18·2
Phenyl magnesium bromide	33·8	34·0

4.8 Hydroxyl compounds

4.8.1 *Alcohols*

Aqueous solutions of methanol and propan-1-ol have been analysed by injecting 50 μl of the sample into 342 g of 80% H_2SO_4, an acid concentration that gave the greatest sensitivity, and the temperature pulse was then referred to a calibration graph. The precision of the method was found to be $\pm0\cdot3\%$ in the analysis of a mixture of propan-1-ol (80·0% v/v) and water (Spink and Spink, 1968).

Alcohols present as impurities in hydrocarbons were determined by Crompton and Cope (1968) who used a continuous-flow apparatus. These and other species (e.g., water and oxygen) that would react with organo-lithium compounds were reacted with 8% triethyl-lithium in the same hydrocarbon. The method has a high sensitivity because of the high enthalpy of the reaction, and contents as low as 1 mg/kg can be determined. The apparatus was operated discontinuously, sampling taking place over a 2-min period every 13 minutes and a Joule heater calibration was interposed every 6 hours over a 20-min period. In the same analysis Russian workers used acetic anhydride as the titrant (Goizman *et al.*, 1971).

Somiya (1930a) determined the acetyl value of oils and fats, using the early titration method. The oil was dried and acetylated with 3 N acetic anhydride, the excess of which was titrated with 3·16 N aniline as described in Section 4.3, and the results given in Table 4.32 were obtained.

Table 4.32 Determination of acetyl value of oils and fats
(Comparison of methods; from Somiya (1932))

Substance (dried)	Thermometric titration	Indicator titration
Castor oil	145·4	142·3
Olive oil	7·1	7
Rape seed oil	7·0	6
Liver oil	6·0	5
Lard	6·1	8

A thermometric investigation into hydrogen-bonding in alcohols was carried out by J. B. Smith (1969).

4.8.2 *Phenols*

Phenols have been determined by utilizing their acidic property and these methods are discussed in Sections 4.1.1(*v*) and 4.1.2(*ii*), but other reactions have also been used. Alkyl phenols have been determined by an acetylation procedure using the direct-injection method (Snelson *et al.*, 1967). In this routine method 1 ml of the sample was injected into 20 ml of an acetylating mixture containing 720 ml of ethyl acetate, 120 ml of acetic anhydride and 4 ml of perchloric acid, the latter acting as a catalyst. The results obtained by this rapid method on a sample of alkyl phenol had a mean hydroxyl content of 1·93% with a repeatability of ±0·07% compared with a repeatability of ±0·10% using a traditional method which after a long acetylation stage requires two titrations, one for the sample and one for the blank.

The reaction of diazonium salts with phenols is rapid and highly exothermic (of the order of −30 kcal/mole) and has been utilized by Pâris and Vial (1952b) in an early titration method for the determination of phenols. In one series, 100 ml of 0·046 M sulphanilic acid diazonium chloride were titrated with 0·5 M solutions of the sodium salts of phenol, o-cresol or 2-naphthol, and in the other series, 100 ml of 0·05 M solution of sodium salts of phenol or 2-naphthol were titrated with 0·45 M p-nitroaniline diazonium chloride. As would be expected from the nature of the reaction the second series was far more accurate. Parsons (1957) used a similar method to determine 2-naphthol.

Polyhydric phenols and phenolic amines are readily oxidized with the formation of quinones and quinimines respectively and this highly exothermic reaction has been utilized by Priestley and his co-workers (1963a) in the evaluation of their automatic digital titrator. They used a 0·25 M cerium(IV) sulphate solution in sulphuric acid as the titrant and the results obtained are given in Table 4.33.

Table 4.33 Determination of phenols by oxidation with cerium(IV)
(From Priestley *et al.* (1963a))

Compound	Concentration	No. of Tests	Precision ±%
Hydroquinone	M/600	8	0·4
Resorcinol	M/480	8	0·8
Metol	M/240	7	0·4
Phenidone	M/600	11	1·8

A back-titration method was also investigated, in which hydroquinone was oxidized with an excess of potassium dichromate in sulphuric acid. The excess of dichromate was then determined by titration with 2 M $Na_2S_2O_3$, an extremely exothermic reaction, so that a 4 mM solution of hydroquinone could be determined with a precision of ±1·4%. This reaction can obviously be applied to a wide range of oxidizable substances.

Thermodynamic data for the reaction of phenol with dimethylformamide were determined by titration of 0·05 M dimethylformamide solution with 1·5 M phenol, both in iso-octane and in toluene solution (Becker *et al.*, 1963a). Using the direct-injection method, Epley and Drago (1967) determined the enthalpy and equilibrium constant of the reaction of phenol with Lewis base in carbon tetrachloride or in cyclohexane solution. In the method, 4·4 M phenol was injected into 110 ml of the solution containing the Lewis base. The bases investigated were acetonitrile, ethyl acetate, acetone, tetrahydrofuran, dimethylacetamide, cyclohexyl chloride, cyclohexyl fluoride, pyridine and triethylamine. A differential direct-injection apparatus has been used to determine the enthalpy of formation of the p-chlorophenol complex with camphor in ether, hexane or carbon tetrachloride solution. The camphor solution was injected into two identical calorimeters, one containing the solvent and the other the p-chlorophenol solution. A value of −6·3 kcal/mole was obtained (Antoine *et al.*, 1968).

The iron(III) complexes of phenol have been investigated and these are discussed in Section 3.36.7.

4.9 Nitriles

A continuous-flow enthalpimetric analyser was used by Taubinger (1969a) to determine free acrylonitrile in large samples of aqueous copolymer latexes. The sample and the reactant (200 g/1 Na_2SO_3) were pumped into the mixing chamber (see Section 2.2.3) through silicone tubing with a peristaltic pump. Regardless of the slowness of the reaction, results on samples containing from 0·05 to 1·0% of free nitrile were correct within $\pm 0·2\%$ absolute but results on polymerization runs containing from 3 to 15% of free nitrile were better, being correct within $\pm 10\%$ relative.

4.10 Olefinic and aromatic compounds

The iodine value of unsaturated fats was determined by Somiya (1930b), using the conventional halogen absorption method. The titrant for this early titration method was prepared by passing chlorine into acetic acid for 2 hours and then standardizing it with oleic acid of known iodine value. In the method a mixture of carbon tetrachloride and acetic acid was used as the solvent. Because the reaction was extremely exothermic, the major part of the titrant was added and the titrand cooled before it was placed in the Dewar flask and the titration completed. In a method proposed by Jordan and his co-workers (1965, 1966) an excess of iodine monochloride was added to the sample (typically a vegetable oil), potassium iodide was added to convert the excess into iodine, which was then titrated with gaseous sulphur dioxide, though the reaction with thiosulphate (see Section 3.29.1) could have been used.

The Diels-Alder reaction of maleic anhydride with dienes has been utilized in an enthalpimetric method for the determination of cyclopentadiene in low-boiling coal-tar oils. In the method, 2·5 g of dry maleic anhydride is added to the dry sample containing between 0·1 and 1·0% of the compound and the temperature rise, which can be as high as 5 degC, is recorded (S.T.P.T.C., 1967).

Benzene has been determined rapidly and with excellent precision in samples containing cyclohexane by measuring the heat of nitration in an enthalpimetric method. A 50-g sample was taken and added to 50 ml of nitration mixture consisting of 2 volumes of 70% nitric acid and 1 volume of 95% sulphuric acid. Calibration standards were prepared containing 0·5–5·0% of benzene in cyclohexane (Willard et al., 1965).

Log K, $\Delta H°$ and $\Delta S°$ values have been determined thermometrically for reactions forming aromatic : salt complexes between Al_2Br_6, $NaAlBr_4$, or $CuAlBr_4$ with benzene or xylene in toluene solvent and for Al_2Br_6 with toluene in benzene solvent. The results which were obtained by using a constant temperature environment calorimeter showed that 1 : 1 and 2 : 1 complexes were formed and indicated that benzene is a stronger electron donor than toluene for these metal salts (Eatough and Van Hecke, 1972).

4.11 Sulphonic acids

2-Naphthalenesulphonic acid can be determined in the presence of 1-naphthalenesulphonic acid and of naphthalenedisulphonic acid by precipitation as the sulphanilamide salt. The reaction is exothermic and has been utilized in a thermometric titration in which sulphanilamide hydrochloride was used as the titrant, and the results obtained were accurate and had a precision of $\pm 1 \cdot 2\%$ (Liebmann *et al.*, 1957). The method, which takes 30 minutes, gave results that compare well with those obtained by the phenylhydrazine precipitation method which takes 3 hours to complete.

4.12 Surfactants

Jordan and his co-workers (1963) stated that well-defined thermometric titration curves could be obtained in the titration of anionic surfactants, a cationic surfactant being used as the titrant. They investigated the titration of 2-dodecylbenzene-1-sulphonate (ABS) with 0·467 M benzyldimethyl(octylphenoxyethoxyethyl)ammonium chloride (Hyamine 1622). This precipitation reaction, in which the hydrophilic groups on the two surfactants are electrostatically neutralized and the solvated water rejected to form the precipitate, is exothermic ($\Delta H = -5 \cdot 8$ kcal/mole). The titration graphs obtained were curved, but the end-point, which was sharp, was preceded by an irregularity attributed to the precipitate adhering to the thermistor. Table 4.34 gives the results obtained and shows

Table 4.34 Precision of methods for the determination of surfactants

Alkylbenzenesulphonate (Jordan *et al.*, 1963) *ABS concentration*, mM		*Precision*	*Cetylpyridinium chloride* (Weiner and Felmeister, 1966). *CPC concentration*, mM		*Precision*
Taken	*Found*	$\pm\%$	*Taken*	*Found*	$\pm\%$
4·00	4·00	0·1	0·500	0·505	—
8·73	8·70	0·5	1·00	1·01	0·02
13·6	13·3	0·6	2·00	2·00	—
17·7	17·7	0·3			

that the method is applicable to solutions of ABS greater than 1 mM in concentration. The precision of the method is far superior to the previous methods, which were time-consuming. Unfortunately it is not applicable to water and sewage analysis where concentrations of ABS of 0·1 mM or less are encountered, unless preliminary enrichment of the sample is carried out.

Weiner and Felmeister (1966) investigated the use of ABS as a titrant in the determination of quaternary ammonium surfactants, but found the titration curve to be rounded as in the previous work. Investigation of other titrants showed that Orange II, an anionic dye (2-hydroxy-1-naphthylazobenzenesulphonate), gave a 1 : 1 insoluble product with solubility products in the range 10^{-10}–10^{-12} at 25 °C and a reaction that was exothermic. In a typical titration, 100 ml of 1 mM cetylpyridinium chloride (CPC) was titrated with 0·108 M Orange II. A sharp end-point was obtained.

4.13 Thio-acids, thio-ethers and disulphides

The enthalpies of the oxidation of the three α-mercaptocarboxylic acids were determined by titration with cerium(IV) in 0·2–1·2 M H_2SO_4 with the results given in Table 4.35. By extrapolation at the end-point, titration volumes were obtained which corresponded to a 1 : 1 Ce : thiol stoichiometry within 1% in all three cases (Alexander *et al.* 1969).

Table 4.35　Heat of oxidation of α-mercaptocarboxylic acids
(From Alexander *et al.* (1969))

Acid	μmole *taken*	$-\Delta H$, kcal/mole
Thiolactic	221·3 147·6	38·4 ± 0·5
Thioglycollic	184·1	40·4 ± 0·6
Thiomalic	173·0 218·4	40·0 ± 0·4

Gol'dshtein and his co-workers have studied the organotin compounds formed with thio-ethers and these are discussed in Section 3.16.4(*ii*).

The heats of formation and the stability constants for the complex formation between iodine and six disulphides have been determined by stepwise addition in ethylene chloride solution. The compounds studied were the methyl, ethyl, propyl, isopropyl, butyl and t-butyl disulphides (Nolander, 1966).

Thiol and thione groups titrated with silver nitrate in propan-2-ol solution have been found to give stable precipitates. In this way mercaptans, thiourea, thioacetamide and tetramethylthiuram disulphide have been determined thermometrically and the errors were found to be between 0·3 and 1·6%. Inflections in the titration curves showed that intermediate compounds were formed and these compounds could often be separated and analysed (Monn, 1966).

Chapter 5

Scope and Application

The reviews of papers given in Chapters 3 and 4 of this book are a collection, as up to date as is possible, of all those so far published, and from them it should be possible to discern the wide fields to which thermometric and enthalpimetric titrimetry have already been applied and the precision of the results obtained. To this end the contents were arranged to make it relatively easy to compare methods of analysis or interactions of particular anions, cations or functional groups. However, variations to these methods should be possible and there is no reason why the more sophisticated modern apparatus should not be substituted for that used in the early titration method. In this way some of the errors that may have arisen in these early procedures because of the large dilution used, which gave rise to stirring problems, combined with the large time-constant of the Beckman thermometer, could possibly be eliminated. Nor is there any reason why a thermometric method cannot be converted into an enthalpimetric method, and where the technique has been used to elucidate a reaction it should be possible to use the same reaction for quantitative analytical purposes.

It is obvious that one of the major uses of the technique is for quantitative analysis and one of the most important questions asked of any technique is what is its precision and accuracy. In the determination of an unknown by thermometric titrimetry using a reaction with an enthalpy change greater than about 15 kcal/mole, results precise and accurate to within $\pm 1\%$ can be obtained in the concentration range of between 1 and 50 mM, an error equivalent to that obtained in a conventional titration at a similar concentration range. At lower concentrations or enthalpies the precision becomes considerably poorer than that obtained by a conventional titration, and precisions of $\pm 1\%$ can only be obtained at the 0·1 mM level if catalysed end-point procedures are used. Titration of a similar concentration of an unknown by an enthalpimetric procedure gives a poorer precision of 1–2% but at the 0·1 mM level a precision of $\pm 5\%$ might still be acceptable, particularly in view of the simplicity and rapidity of this technique. Precisions of $\pm 2\%$ have been reported for continuous-flow methods.

Another use of the technique is in the determination of the thermo-dynamic parameters of a reaction. It has been stated that 'thermometric titrimetry is particularly well suited for the determination of $\Delta H°$, $\Delta S°$ and $\Delta G°$ (and thus pK) from a single titration since in a single run, one obtains the equivalent of a large number of determinations by conventional calorimetry' (Christensen *et al.*, 1966a). This view has been criticized by Cabani and Gianni (1972) who advise that the observed heat change must be several orders greater than the sensitivity of the apparatus used and that it is important, in order to avoid systematic errors, that different concentrations of both titrant and titrand should be investigated as well as the reversal of the rôles of the reactants used for the titrant and titrand. The precision expected in these sophisticated titrations is of the order of 0·03 kcal/mole for $\Delta H°$ and 0·05 for pK. However, data of this kind are still limited, and if precisions of the order of 2% or greater can be tolerated then with more simple apparatus and techniques, thermometric (or even enthalpimetric) titrimetry is an excellent method of establishing whether suspect earlier calorimetric data need to be checked.

The advantages of the technique considerably outnumber its disadvan-tages, the most outstanding being the universal nature of the detector used in the titrations. It is not affected by the colour of the solution, as in the case of indicator titrations, nor by the reactivity of the solutions involved that often inactivate glass or other electrodes, especially in non-aqueous solution. Unusual media, such as fused salts, or titrants, such as hydro-fluoric acid, can thus be used. As is typical of a titration technique it has the advantage of speed over a gravimetric procedure but unlike indicator or potentiometric titrations can also be used in many systems that produce precipitates or even gels. Its use can often avoid the difficulty that can arise in trying to find a suitable indicator or electrode system that does not, for example, interact with the reactants, the products or in particular, the solvent. Though less precise, direct-injection enthalpimetry has the added advantage of speed of analysis and the attraction of the use of non-standardized titrants coupled with the possibility of multiple serial analysis as exemplified by the work of Sajó and Sipos. This technique can also be carried out on a miniature scale, a feature which has not yet been fully exploited, and can also be used for plant control. The latter can, however, be much more readily carried out by continuous-flow enthalpi-metry.

A restriction on the wide use of the technique has been caused by the lack of availability of commercially produced apparatus and consequently many workers have had to construct their own. On the whole the dis-advantages encountered are minor and are principally concerned with the avoidance of extraneous heat effects as discussed in Section 2.6. It has always to be remembered that the technique is concerned with heat, and foresight is needed to avoid unnecessary difficulties. A large volume of a solution may not have an even temperature, a solution brought from another part of the laboratory may be at a different temperature, heat loss may occur by evaporation because of too vigorous stirring, there is need to take excessive care (and thus time) to equalize the titrant and titrand temperatures when the heat change involved is small; these are all examples of these difficulties.

Those wishing to develop a thermometric or enthalpimetric method of analysis will see from the theoretical aspects discussed in Chapter 1 that this is only practicable if the following conditions can be satisfied with respect to the reaction being studied or utilized.

(i) Single-component systems

(a) The heat change is large enough to produce a significant inflection in a thermometric titration curve or a large enough heat pulse in an enthalpimetric titration. If this is not possible, either because the concentration of the unknown is too low or the heat of the reaction is too small (see Table 1.1), then use of end-point enhancement techniques based on secondary reactions, catalytic effects, heats of dilution or other effects should be considered.

(b) The free energy change or its significant component, the equilibrium constant, K, should be such that the curvature of a thermometric titration curve at the end-point is not so great that back-extrapolation cannot be used.

(c) The kinetics are faster than the rate of addition or generation of the titrant. If the titration rate in a slow-reacting system is impractically slow, the use of a reaction catalyst should be considered.

(ii) Multi-component or multi-ionization systems

As well as the conditions above the following must also be satisfied with respect to a thermometric titration.

(a) The heat of reaction of one of two consecutively titrated species should be 50% greater than that of the other so that the titration curve will show a sufficient inflection to mark the intermediate end-point in the titration.

(b) The difference between the equilibrium constants of the two consecutively titrated species must be sufficient for the first to be titrated to virtual completion before the titration of the second commences unless the 'total moles–total enthalpy' method (see Section 1.5.6) can be used.

The choice of apparatus to be used to carry out the analysis will depend primarily on the heat of reaction; the greater this is at a particular concentration and precision required, the more simple the apparatus required. This is the reason for the superiority of the catalysed end-point methods as the heat-change is independent of the reaction involved in the determination. Where the heat of the reaction is not known the use of an apparatus consisting of a beaker surrounded by expanded polystyrene with the simplest of motorized syringes and thermistor detectors can provide sufficient information as to the feasibility of a proposed method. A choice can then be made, based on this information and the precision required, as to the final form of the apparatus.

The following tabulation summarizes the applicability of this versatile analytical technique (after Jordan, 1963).

Titration procedure	*Information obtainable*	*Method*
Enthalpimetric	Concentration of an unknown	Calibration
	Heat of reaction, ΔH	Calibration
Thermometric	Amount of an unknown	Calibration
	Heat of reaction, ΔH	Initial slope of titration curve
	Free energy of reaction, ΔG	Curvature of titration curve at end-point
	Reaction stoichiometry	End-point determination
	Reaction entropy	By calculation from ΔH and ΔG

References

ACKERMANN, T. *Z. Electrochem.* **62**, 411 (1958).

ALEXANDER, W. A., MASH, C. J., MCAULEY, A. *Talanta* **16**, 535–539 (1969).

ALEXANDER, W. A., MASH, C. J., MCAULEY, A. *Analyst* **95**, 657–660 (1970).

ALLEMAN, T. G. Abstr. of Papers, 132nd Meeting, Sept., *Am. Chem. Soc.*, p. 11B (1957).

ALLEN, K. E. *Anal. Chem.* **28**, 277 (1956).

ANDERSON, K. P., NEWELL, D. A., IZATT, R. M. *Inorg. Chem.* **5**, 62–65 (1966a).

ANDERSON, K. P., GREENHALGH, W. O., IZATT, R. M. *Inorg. Chem.* **5**, 2106–2124 (1966b).

ANDERSON, K. P., GREENHALGH, W. O., IZATT, R. M. *Inorg. Chem.* **6**, 1056–1058 (1967).

ANTOINE, B., LAURANSON, J., SAUMAGNE, P., BAUMGARTNER, P. *Rev. Inst. Fr. Petrole Ann. Combust Liquides* **23**, 1389–1396 (1968).

ARNEK, R. *Arkiv. Kemi* **24**, 531–549 (1965).

ASHMORE, S. A., THICKIN, D. *Coke & Gas* **11**, 307–308 (1949).

AVEDIKIAN, L. *Bull. Soc. Chim. France* 2570–2576 (1966).

AVEDIKIAN, L. *Bull. Soc. Chim. France* 254–255 (1967).

BANERJEE, S. *Sci. Cult., Calcutta* **16**, 115 (1950a).

BANERJEE, S. *J. Indian Chem. Soc.* **27**, 417–424 (1950b).

BANERJEE, S., HALDAR, B. C. *Nature* **165**, 1012 (1950).

BANERJEE, S., MITRA, S. K. *Sci. Cult., Calcutta* **16**, 530–531 (1951).

BARK, L. S., BARK, S. M. 'Thermometric Titrimetry', Pergamon Press, Oxford (1969).

BARK, L. S., BATE, P. *Analyst*, **97**, 783–786 (1972).

BARK, L. S., DORAN, K. Quoted by Bark and Bark (1969).

BARK, L. S., GRIME, J. K. Society for Analytical Chemistry Meeting, Swansea, May 17th (1972).

BARRON, S., RAUTIO, W. S. *Instr. Control Systems* **41**, 83–85 (1968).

BARTHEL, J., SCHMAHL, N. G. *Z. Anal. Chem.* **207**, 81–90 (1965).

BARTHEL, J., BECKER, F., SCHMAHL, N. G. *Z. Physik. Chem.* (*Frankfurt*), **29**, 57–76 (1961).

BARTHEL, J., SCHMAHL, N. G., LENZ, K. *Z. Anal. Chem.* **233**, 328–340 (1968).

BECK, A. *J. Sci. Instr.* **33**, 16–17 (1956).

BECKER, F., GRUNDMANN, R. *Z. Physik. Chem.* (*Frankfurt*) **66**, 137–149 (1969).

BECKER, F., LÜSCHOW, H. M. *Proc. Int. Conf. Coord. Chem.*, 8th, Vienna, p. 334–338 (1964).

BECKER, J. A., GREEN, C. B., PEARSON, G. L. *Bell System Tech. J.*, **26**, 170–211 (1947).

BECKER, F., BARTHEL, J., SCHMAHL, N. G. *Z. Physik. Chem.* (*Frankfurt*) **37**, 33–51 (1963a).

BECKER, F., BARTHEL, J., SCHMAHL, N. G., LÜSCHOW, H. M. *Z. Physik. Chem.* (*Frankfurt*) **37**, 52–62 (1963b).

BEEZER, A. E. Paper given at *Soc. Anal. Conf.*, Durham (1970).

BEEZER, A. E., SLAWINSKI, A. K. *Talanta* **18**, 837 (1971).

BELCHER, R., TATLOW, J. C. *Analyst* **76**, 593–595 (1951).

BELCHER, R., TULLY, G. W., SVEHLA, G. *Anal. Chim. Acta* **50**, 261–267 (1970).

BELISLE, J. *Anal. Chim. Acta* **54**, 156–158 (1971).

BELL, J. M., COWELL, C. F. *J. Am. Chem. Soc.* **35**, 49–54 (1913).

BENET, L. Z. *Diss. Abstr.* **27B**, 1434 (1966).

BEN-YAIR, M. P. Abstr. of Papers, 132nd Meeting, Sept., *Am. Chem. Soc.* p. 6B (1957).

BEN-YAIR, M. P. *Trans. Chalmers Univ. Technol.*, Gothenburg, No. 236, 3–8 (1961).

BEN-YAIR, M. P., JORDAN, J. Abstr. of Papers, XIIth International Congress of Pure and Applied Chemistry, N.Y. Sept. 42–43 (1951).

BETHGE, P. O. *Anal. Chim. Acta* **9**, 129–139 (1953).

BHADRAUER, M. S., GUAR, J. N. *J. Indian Chem. Soc.* **36**, 103–107 (1959).

BHATTACHARYA, A. K., GUAR, H. C. *J. Indian Chem. Soc.* **24**, 487–494 (1947).

BHATTACHARYA, A. K., GAUR, H. C. *J. Indian Chem. Soc.* **25**, 183–190 (1948).

BHATTACHARYA, A. K., SAXENA, R. S. *J. Indian Chem. Soc.* **29**, 263–269 (1952a)

BHATTACHARYA, A. K., SAXENA, R. S. *J. Indian Chem. Soc.* **29**, 529–534 (1952b).

BILLINGHAM, E. J., Jr., REED, A. H. *Anal. Chem.* **36**, 1148–1149 (1964).

BJORKMAN, M., SILLÉN, L. G. *Trans. Roy. Inst. Technol. Stockholm*, No. 199, 18 pp. (1963).

BOBTELSKY, M., JORDAN, J. *J. Am. Chem. Soc.* **69**, 2286–2290 (1947).

BONNER, T. G. *Analyst* **71**, 483–492 (1946).

BOSE, M., CHOWDHURY, D. W. *J. Indian Chem. Soc.* **32**, 673–678 (1955).

BOYD, S., BRYSON, A., NANCOLLAS, G. H., TORRANCE, K. *J. Chem. Soc.*, 7353–7358 (1965).

BRENNER, A. *J. Electrochem. Soc.* **112**, 611–621 (1965).

BRESCIA, J., PEISACH, F. *J. Am. Chem. Soc.* **76**, 5946–5948 (1954).

British Standard 2087: 1971, Preservative Textile Treatments, 38–39.

BROWN, E. G., HAYES, T. J. *Anal. Chim. Acta* **9**, 1–5 (1953).

BROWN, M. W., ISSA, K., SINCLAIR, A. G. *Analyst* **94**, 234–235 (1969).

BRUNETTI, A. P., LIM, M. C., NANCOLLAS, G. H. *J. Am. Chem. Soc.* **90,** 5120–5126 (1968).

BRUNETTI, A. P., NANCOLLAS, G. H., SMITH, P. N. *J. Am. Chem. Soc.* **91,** 4680–4683 (1969).

BURTON, K. C., IRVING, H. M. N. H. *Anal. Chim. Acta.* **52,** 441–446 (1970).

CABANI, S., GIANNI, P. *Anal. Chem.* **44,** 253–259 (1972).

CADARIV, I., GOINA, T. *Studia Univ. Babes-Bolyai, Ser. 1,* 25–35 (1961).

CARE, R. A., STAVELEY, L. A. R. *J. Chem. Soc.* 4571–4579 (1956).

CARR, P. W. *Anal. Chem.* **43,** 756–758 (1971a).

CARR, P. W. *Thermochimica Acta* **2,** 505–511 (1971b).

CARR, P. W. *Thermochimica Acta* **3,** 427–435 (1972).

CHARLES, R. G. *J. Am. Chem. Soc.* **76,** 5854–5858 (1954).

CHATTERJI, K. K. *Sci. Cult. (Calcutta)* **14,** 530 (1949).

CHATTERJI, K. K. *J. Indian Chem. Soc.* **32,** 366–370 (1955).

CHATTERJI, K. K. *J. Indian Chem. Soc.* **35,** 57–62 (1958a).

CHATTERJI, K. K. *J. Indian Chem. Soc.* **35,** 709–711 (1958b).

CHATTERJI, K. K. *J. Indian Chem. Soc.* **35,** 883–889 (1958c).

CHATTERJI, K. K., GHOSH, A. K. *J. Indian Chem. Soc.* **34,** 407–412 (1957).

CHIPPERFIELD, J. R. *Proc. Roy. Soc. (London), Ser. B* **164,** 401–410 (1966).

CHIPPERFIELD, J. R., ROSSI-BERNARDI, L., ROUGHTON, F. J. W. *J. Biol. Chem.* **242,** 777–783 (1967).

CHRISTENSEN, J. J., IZATT, R. M. *J. Phys. Chem.* **66,** 1030–1034 (1962).

CHRISTENSEN, J. J., IZATT, R. M., HANSEN, L. D. *Rev. Sci. Instr.* **36,** 779–783 (1965).

CHRISTENSEN, J. J., IZATT, R. M., HANSEN, L. D., PARTRIDGE, J. A. *J. Phys. Chem.* **70,** 2003–2010 (1966a).

CHRISTENSEN, J. J., RYTTING, J. H., IZATT, R. M. *J. Am. Chem. Soc.* **88,** 5105–5106 (1966b).

CHRISTENSEN, J. J., RYTTING, J. H., IZATT, R. M. *J. Phys. Chem.* **71,** 2700–2705 (1967).

CHRISTENSEN, J. J., JOHNSTON, H. D., IZATT, R. M. *Rev. Sci. Instr.* **39,** 1356–1359 (1968a).

CHRISTENSEN, J. J., WRATHALL, D. P., IZATT, R. M. *Anal. Chem.* **40,** 175–181 (1968b).

CHRISTENSEN, J. J., WRATHALL, D. P., OSCARSON, J. O., IZATT, R. M. *Anal. Chem.* **40,** 1713–1717 (1968c).

CHRISTENSEN, J. J., RYTTING, J. H., IZATT, R. M. *J. Chem. Soc. A,* 861–862 (1969).

CIOFFI, F. J., ZENCHELSKY, S. T. *J. Phys. Chem.* **67,** 357–359 (1963); CIOFFI, F. J. *Diss. Abstr.,* **23,** 46–47 (1962).

COBB, W. D., HARRISON, T. S. *Analyst* **99,** 764–770 (1971).

CRESPIN, G. *Lapis,* No. 71, 24 (1964).

CROMPTON, T. R., COPE, B. *Anal. Chem.* **40,** 274–280 (1968).

DAFTARY, R. D., HALDAR, B. C. *Anal. Chim. Acta* **25,** 538–541 (1961).

DANEST, Z., TRISCHLER, F. *Z. Anal. Chem.* **256,** 129 (1971).

DANIELSSON, I., NELANDER, B., SUNNER, S., WADSÖ, I. *Acta Chem. Scand.* **18,** 995–998 (1964).

DANIELSSON, I., INGMAN, L., SCHALIEN, R. V. *Suomen Kemistilehti*, **38B**, 43–48 (1965).

DASTOOR, M. J., HALDAR, B. C. *Indian J. Chem.* **5**, 335–337 (1967).

DEAN, P. M., NEWCOMER, E. *J. Am. Chem. Soc.* **47**, 64–67 (1925).

DEAN, P. M., WATTS, O. O. *J. Am. Chem. Soc.* **46**, 855–858 (1924).

DE CARVALHO, R. G., CHOPPIN, G. R. *J. Inorg. Nucl. Chem.* **29**, 737–743 (1967).

DEGISCHER, G., NANCOLLAS, G. H. *Inorg. Chem.* **9**, 1259–1262 (1970).

DE LEO, A. B., STERN, M. J. *J. Pharm. Sci.* **53**, 993–994 (1964).

DE LEO, A. B., STERN, M. J. *J. Pharm. Sci.* **54**, 911–914 (1965).

DE LEO, A. B., STERN, M. J. *J. Pharm. Sci.* **55**, 173–180 (1966).

DESCHAMPS, P., DEBURCK, A., BONNAIRE, Y. *Anal. Chim. Acta* **40**, 259–266 (1968).

DILKE, M. K., ELEY, D. D. *J. Chem. Soc.* 2601–2612; 2613–2615 (1949).

DOBKIN, V. M., BATRAKOVA, N. S., GERULAĬTIS, YU. N., GERMAN, V. N., KURSKAYA, M. L., OSTROVSKII, YU. I. U.S.S.R. Patent 138 088 (1961).

DOLINSKIĬ, P. I. *Zavodsk. Lab.* **6**, 553–555 (1937).

DOLINSKIĬ, P. I. *Zavodsk. Lab.* **7**, 223–226 (1938).

DRAGO, R. S., BOLLES, T. F., NEIDZIELSKI, R. J. *J. Am. Chem. Soc.* **88**, 2717–2721 (1966).

DRĂGULESCU, C., POLICEC, S. *Acad. Rep. Populare Romine. Baza Cerc. Stii. Tim. Studii Cercetari Chim.* **9**, 33–40 (1962).

DUMBAUGH, W. H., Jr. *Diss. Abstr.* **20**, 1559 (1959).

DUTOIT, P., GROBET, É. *J. Chim. Phys.* **19**, 324–327 (1921).

EATOUGH, D. J. *Anal. Chem.* **42**, 635–639 (1970).

EATOUGH, D. J., VAN HECKE, G. R. *Thermochimica Acta*, **3**, 165–170 (1972).

EPLEY, T. D., DRAGO, R. S. *J. Am. Chem. Soc.* **89**, 5770–5773 (1967).

ERDEY, L., MARIK, J. *Magy. Kem. Lapja* **25**, 584–585 (1970).

EVERETT, D. H., HYNE, J. B. *J. Chem. Soc.* 1636–1642 (1958).

EVERETT, D. H., WYNNE-JONES, W. F. K. *Trans. Faraday. Soc.* **35**, 1380 (1939).

EVERSON, W. L. *Anal. Chem.* **36**, 854–856 (1964).

EVERSON, W. L. *Anal. Chem.* **39**, 1894–1896 (1967).

EVERSON, W. L. *Anal. Chem.* **43**, 201–205 (1971).

EVERSON, W. L., RAMIREZ, E. M. *Anal. Chem.* **37**, 806–811 (1965a).

EVERSON, W. L., RAMIREZ, E. M. *Anal. Chem.* **37**, 812–816 (1965b).

EVERSON, W. L., RAMIREZ, E. M. *Anal. Chem.* **39**, 1771–1776 (1967).

EWING, G. J. *Diss. Abstr.* **21**, 1731–1732 (1961).

EWING, G. J., MAZAC, C. J. *Anal. Chem.* **38**, 1575–1577 (1966).

FAY, D. P., PURDIE, N. *J. Phys. Chem.* **73**, 3462–3467 (1969).

FLENGAS, S. N., RIDEAL, E. *Proc. Roy. Soc. A.* **233**, 443 (1956).

FORMAN, E. J., HUME, D. N. *J. Phys. Chem.* **63**, 1949–1952 (1959).

FORMAN, E. J., HUME, D. N. *Talanta* **11**, 129–137 (1964).

FREEBERG, F. E. *Anal. Chem.* **41**, 54–57 (1969).

FRIEDMAN, H. G., Jr. *Diss. Abstr.* **27B**, 1076 (1966).

FRITZ, J. S. *Anal. Chem.* **25**, 407–411 (1953).

GALLET, J.-P., PÂRIS, R. A. *Anal. Chim. Acta* **39**, 181–188 (1967a).

GALLET, J.-P., PÂRIS, R. A. *Anal. Chim. Acta* **39**, 341–348 (1967b).

GAUR, H. C., BHATTACHARYA, A. K. *J. Indian Chem. Soc.* **26**, 46–52 (1949).

GAUR, H. C., BHATTACHARYA, A. K. *Proc. Natl. Acad. Sci., India* **19A**, 45–53 (1950a).

GAUR, H. C., BHATTACHARYA, A. K. *J. Indian Chem. Soc.* **27**, 131–137 (1950b).

GAUR, H. C., BHATTACHARYA, A. K. *J. Indian Chem. Soc.* **29**, 29–33 (1952a)

GAUR, H. C., BHATTACHARYA, A. K. *J. Indian Chem. Soc.* **29**, 117–122 (1952b).

GAUR, J. N., BHADRAUER, M. S. *J. Indian Chem. Soc.* **36**, 108–110 (1959).

GAUR, J. N., GAUR, H. C., BHATTACHARYA, A. K., *J. Indian Chem. Soc.* **30**, 859–862 (1953).

GAUR, J. N., GAUR, H. C., BHATTACHARYA, A. K. *J. Indian Chem. Soc.* **35**, 144–146 (1958).

GERDING, P. *Acta Chem. Scand.* **20**, 79–94 (1966a).

GERDING, P. *Acta Chem. Scand.* **20**, 2771–2780 (1966b).

GERDING, P. *Acta Chem. Scand.* **23**, 1695–1703 (1969).

GERDING, P., LÉDEN, I., SUNNER, S. *Acta Chem. Scand.* **17**, 2190–2198 (1963).

GLEMSER, O., EINERHAND, J. *Z. Anorg. Allgem. Chem.* **261**, 26–42 (1950).

GODIN, M. C. *J. Sci. Instr.* **39**, 241–242 (1962).

GOIZMAN, M. S. *Dokl. Akad. Nauk, SSSR* **184**, 599–601 (1969).

GOIZMAN, M. S., IVANOVA, I. L., BANDAREVA, T. I. *Khim. Farm. Zh.* **5**, 55–57 (1971).

GOL'DSHTEIN, I. P., GUR'YANOVA, E. N., KOCHESHKOV, K. A. *Dokl. Akad. Nauk SSSR* **161**, 111–114 (1965a).

GOL'DSHTEIN, I. P., GUR'YANOVA, E. N., KARPOVITCH, I. R. *Zh. Fiz. Khim.* **39**, 932–937 (1965b).

GOL'DSHTEIN, I. P., GUR'YANOVA, E. N., ZEMLYANSKII, N. N., SYUTKINA, O. P., PANOV, E. M., KOCHESHKOV, K. A. *Dokl. Akad. Nauk SSSR* **175**, 836–839 (1967a).

GOL'DSHTEIN, I. P., GUR'YANOVA, E. N., ZEMLYANSKII, N. N., SYUTKINA, O. P., PANOV, E. M., KOCHESHKOV, K. A. *Izv. Akad. Nauk SSSR Ser. Khim.* 2201–2207 (1967b).

GOUDARD, L., GRANGETTO, A. *Compt. Rend.* **263**, 10–12 (1966).

GOYAN, F. M., JOHNSON, R. D., BLACKWOOD, R. H. *J. Am. Pharm. Assoc.* **50**, 773–777 (1961).

GRAY, V. R., WHELAN, P. F. *Chem. Ind. (London)* 126–128 (1955).

GREATHOUSE, L. H. Abstr. of Papers, Sept. Meeting *Am. Chem. Soc.* p. 7B (1957).

GREATHOUSE, L. H., JANSSEN, H. J., HAYDEL, C. H. *Anal. Chem.* **28**, 357–361 (1956).

GREENHOW, E. J., SPENCER, L. E. *Chem. Ind. (London)* 422 (1972a).

GREENHOW, E. J., SPENCER, L. E. *Chem. Ind. (London)* 466 (1972b).

GREENHOW, E. J., SPENCER, L. E. Private communication; *Analyst*, **98**, 81, 90, 98 (1973).

GRENTHE, I. *Acta Chem. Scand.* **18**, 283–292 (1964).

GRENTHE, I., WILLIAMS, D. R. *Acta Chem. Scand.* **21**, 347–358 (1967).

GRENTHE, I., OTS, H., GINSTRUP, O. *Acta Chem. Scand.* **24,** 1067–1080 (1970).

GROBET, É. *J. Chim. Chys.* **19,** 331–335 (1921).

GUÉRIN, H. *Bull. Soc. Chim. France* **19,** 24–35 (1952).

GUILLOT, P. *Anal. Chim. Acta,* **50,** 499–504 (1970).

HALDAR, B. C. *J. Indian Chem. Soc.* **23,** 147–152 (1946a).

HALDAR, B. C. *J. Indian Chem. Soc.* **23,** 153–156 (1946b).

HALDAR, B. C. *J. Indian Chem. Soc.* **23,** 183–188 (1946c).

HALDAR, B. C. *J. Indian Chem. Soc.* **23,** 205–210 (1946d).

HALDAR, B. C. *J. Indian Chem. Soc.* **24,** 503–510 (1947).

HALDAR, B. C. *J. Indian Chem. Soc.* **25,** 439 (1948a).

HALDAR, B. C. *J. Indian Chem. Soc.* **25,** 439–442 (1948b).

HALDAR, B. C. *J. Indian Chem. Soc.* **25,** 445–447 (1948c).

HALDAR, B. C. *Nature* **166,** 744–745 (1950a).

HALDAR, B. C. *J. Indian Chem. Soc.* **27,** 484–492 (1950b).

HALE, J. D., IZATT, R. M., CHRISTENSEN, J. J. *J. Phys. Chem.* **67,** 2605–2612 (1963).

HALL, N. F. *J. Am. Chem. Soc.* **52,** 5115–5128 (1930).

HALLET, L. T. *Anal. Chem.* **24,** 1348 (1952).

HANSEN, L. D. *Diss. Abstr.* **26,** 5000 (1966).

HANSEN, L. D., LEWIS, E. A. *J. Chem. Thermodynamics* **3,** 35–41 (1971b).

HANSEN, L. D., CHRISTENSEN, J. J., IZATT, R. M. *Chem. Commun.* **3,** 36–38 (1965).

HANSEN, L. D., LEWIS, E. A. *Anal. Chem.* **43,** 1393–1397 (1971a).

HARNED, H. S., ROBINSON, R. A. *Trans. Faraday Soc.* **36,** 973 (1940).

HARRIES, R. J. N. *Proc. Soc. Anal. Chem.* **3,** 19–20 (1966).

HARRIES, R. J. N. *Talanta* **15,** 1345–1352 (1968).

HARRIS, P. C., MOORE, T. E. *Inorg. Chem.* **7,** 656–660 (1968).

HEUBLEIN, G., RAUSCHER, B. *Tetrahedron* **25,** 3999–4004 (1969).

HILDEBRAND, J. H., SCOTT, R. L. 'Regular Solutions', Prentice Hall, New Jersey (1962).

HOFFMANN, E. G., TORNAU, W. Z. *Anal. Chem.* **188,** 321–334 (1962).

HOLMES, F., WILLIAMS, D. R. *J. Chem. Soc., A* 729–731 (1967a).

HOLMES, F., WILLIAMS, D. R. *J. Chem. Soc. A* 1256–1259 (1967b).

HOLMES, F., WILLIAMS, D. R. *J. Chem. Soc. A* 1702–1704 (1967c).

HOWARD, H. *J. Soc. Chem. Ind.* (*London*) **29,** 3–4 (1910).

HUME, D. N., DUFFIELD, B. J. Abstr. of Papers, 147th Meeting, *Am. Chem. Soc., Philadelphia, Pa.* p. 13B (1964).

HUME, D. N., JORDAN, J. *Anal. Chem.,* **30,** 2064–2065 (1958).

IZATT, R. M., CHRISTENSEN, J. J., PACK, R. T., BENCH, R. *Inorg. Chem.* **1,** 828–831 (1962).

IZATT, R. M., HANSEN, L. D., RYTTING, J. H., CHRISTENSEN, J. J. *J. Am. Chem. Soc.* **87,** 2760–2761 (1965).

IZATT, R. M., RYTTING, J. H., HANSEN, L. D., CHRISTENSEN, J. J. *J. Am. Chem. Soc.* **88,** 2641–2645 (1966).

IZATT, R. M., JOHNSTON, H. D., WATT, G. D., CHRISTENSEN, J. J. *Inorg. Chem.* **6,** 132–135 (1967).

IZATT, R. M., EATOUGH, D. J., SNOW, R. L., CHRISTENSEN, J. J. *J. Phys. Chem.* **72,** 1208–1213 (1968).

IZATT, R. M., EATOUGH, D. J., CHRISTENSEN, J. J., BARTHOLOMEW, C. H. *J. Chem. Soc. A* 45–47 (1969).

IZATT, R. M., JOHNSTON, H. D., EATOUGH, D. J., HANSEN, J. W., CHRISTENSEN, J. J. *Thermochimica Acta* **2,** 77–85 (1971).

JAHR, K. F., WIESE, G., SCHUCHARDT, G. *Z. Physik. Chem. (Frankfurt),* **61,** 73–91 (1968).

JAMBON, C., MERLIN, J.-C. *Compt. Rend.* **272C,** 195–197 (1971).

JESPERSEN, N. D., JORDAN, J. *Anal. Lett.* **3,** 323–334 (1970).

JOHANSSON, C. E. *Talanta* **17,** 739–745 (1970).

JOHANSSON, S. *Arkiv Kemi* **24,** 189–209 (1965).

JORDAN, J. *J. Chem. Educ.* **40,** A5–A21 (1963a).

JORDAN, J. *Chimia Aarau* **17,** 101–136 (1963b).

JORDAN, J. 'Thermometric and Enthalpy Titrations', in 'Treatise on Analytical Chemistry', (I. M. Kolthoff and P. Elving, eds.) Part 1, Theory and Practice Vol. 8, Interscience, N.Y. pp. 5175–5242 (1968).

JORDAN, J., ALLEMAN, T. G. *Anal. Chem.* **29,** 9–13 (1957).

JORDAN, J., BEN-YAIR, M. P. *Arkiv Kemi* **11,** 239–249 (1957).

JORDAN, J., BILLINGHAM, E. J., Jr. *Anal. Chem.* **33,** 120–123 (1961).

JORDAN, J., DUMBAUGH, W. H., Jr. *Anal. Chem.* **31,** 210–213; Abstr. of Papers, 132nd Meeting, *Am Chem. Soc.* N.Y. Sept. 1957, p. 11B (1959a).

JORDAN, J., DUMBAUGH, W. H., Jr. *Bull. Chem. Thermo. (IUPAC)* **2,** *Sect. A–15,* 9–11 (1959b).

JORDAN, J., EWING, G. J. *Bull. Chem. Thermo. (IUPAC)* **3,** *Sect. A–13*(a), 7–8 (1960).

JORDAN, J., MEIER, J., BILLINGHAM, E. J., Jr., PENDERGRAST, J. *Chem. Progr.* **19,** 193 (1958).

JORDAN, J., MEIER, J., BILLINGHAM, E. J., Jr., PENDERGRAST, J. *Anal. Chem.* **31,** 1439–1446 (1959a).

JORDAN, J., MEIER, J., BILLINGHAM, E. J., Jr., PENDERGRAST, J. *Bull. Chem. Thermo. (IUPAC)* **2,** *Sect. A–15,* 11 (1959b).

JORDAN, J., MEIER, J., BILLINGHAM, E. J., Jr., PENDERGRAST, J. *Anal. Chem.* **32,** 651–655 (1960a).

JORDAN, J., MEIER, J., BILLINGHAM, E. J., Jr., PENDERGRAST, J. *Nature* **187,** 318–319 (1960b).

JORDAN, J., MEIER, J., BILLINGHAM, E. J., Jr., PENDERGRAST, J. *Bull. Chem. Thermo. (IUPAC)* **3,** Sect. A–13 (b), 8–9 (1960c).

JORDAN, J., PEI, P. T., JAVICK, R. A. *Anal. Chem.* **35,** 1534 (1963).

JORDAN, J., PEI, P. T., BUCHTA, R. C. Abstr. of Papers, 149th Meeting, *Am. Chem. Soc.,* p. 30B–31B (1965).

JORDAN, J., HENRY, R. A., WASILEWSKI, J. C. *Microchem. J.* **10,** 260–272 (1966).

KEILY, H. J., HUME, D. N. *Anal. Chem.* **28,** 1294–1297 (1956).

KEILY, H. J., HUME, D. N. *Anal. Chem.* **36,** 543 (1964); HUME, D. N., KEILY, H. J. Abstr. of Papers, 132nd Meeting, *Am. Chem. Soc.* p. 6B (1957).

KETTRUP, A., SPECKER, H. *Z. Anal. Chem.* **230,** 241–250 (1967).

KHOLLER, V. A. *Vestn. Mosk. Univ. Khim.* **24,** 120–121 (1969).

KISS, T. A. *Z. Anal. Chem.* **252,** 12–13 (1970).

KORENMAN, I. M., SHEYANOVA, F. R., NIKOLAEV, B. A., ABRAMOV, O. B. *Tr. po. Khim. i Khim. Tekhnol.* **4,** 753–760 (1961).

KORTÜM, G., VOGEL, W., ANDRUSSOW, K. *Pure Appl. Chem.* **1,** 224 (1961).

KUGLER, G. C., CAREY, G. H. *Talanta* **17,** 907–914 (1970).

LANDOLT-BÖRNSTEIN 'Physikalische-chemische Tabellen', Springer, Berlin (1923).

LÉDEN, I., RYHL, T. *Acta Chem. Scand.* **18,** 1196–1201 (1964).

LEDESMA, R. REYNOLDS, C. A. Abstr. of Papers, 155th Meeting, *Am. Chem. Soc.* p. B 85 (1968).

LEIBMANN, W., PARSONS, J. S., WOODS, J. T. Abstr. of Papers, 132nd Meeting, *Am. Chem. Soc.,* N.Y., p. 12B (1957).

LEVENE, P. A., BASS, L. W., SIMMS, H. S. *J. Biol. Chem.* **70,** 229; 243 (1926).

LEWIS, C. D. Abstr. of Papers, 132nd Meeting, *Am. Chem. Soc.* N.Y., p. 7B (1957).

LEYDET, P., CHAMPETIER, G. *Compt. Rend.* **262C,** 48–51 (1966).

LINDE, H. W., ROGERS, L. B., HUME, D. N. *Anal. Chem.* **25,** 404–407 (1953).

LINGANE, J. J. *Anal. Chem.* **20,** 285–292 (1948).

LOPACHAK, G. G., GUDZ, L. F. *Ogneupary,* **36,** 65–69 (1971).

MCCLURE, J. H., RODER, T. M., KINSEY, R. H. *Anal. Chem.* **27,** 1599–1601 (1955).

MACKENZIE, R. C. *Talanta* **16,** 1227–1230 (1969).

MCLEAN, W. R., PENKETH, G. E. *Talanta* **15,** 1185–1197 (1968).

MALINGER, M., BANSTETOR, J., HULEGA, J. *Chem. Listy* **63,** 931–938 (1969).

MANDL, M., SKALA, J., KASE, M. *Neue Hütte* **7,** 176–180 (1962).

MARIK-KORDA, P., ERDEY, L. *Talanta* **17,** 1215–1218 (1970).

MARINENKO, G., TAYLOR, J. K. *J. Res. Natl. Bur. Stds.* **70C,** 1–3 (1966).

MAYR, C., FISCH, J. *Z. Anal. Chem.* **76,** 418–438 (1929).

MEAD, T. E. *J. Phys. Chem.* **66,** 2149–2154 (1962).

MEITES, T., MEITES, L., JAITLY, J. N. *J. Phys. Chem.* **73,** 3801–3809 (1969).

MELLON, M. G. 'Qualitative Analysis', Gowell, New York, p. 555 (1955).

MERÉNYI, G. *Nehézvegyip. Kut Int. Kozlemen.* **3,** 139–143 (1966).

METZGER, W. H., Jr., BRENNER, A., SALMAN, H. I. *J. Electrochem. Soc.* **114,** 131–138 (1967).

MILLER, F. J., THOMASON, P. F. *Anal. Chem.* **31,** 1498–1500 (1959a).

MILLER, F. J., THOMASON, P. F. *Anal. Chim. Acta* **21,** 112–116 (1959b).

MILLER, F. J., THOMASON, P. F. *Talanta* **2,** 109–114 (1959c).

MISUMI, S., TAKETATSU, T. *Bull. Soc. Chim. Japan* **32,** 6 (1959).

MONDAIN–MONVAL, P., PÂRIS, R. *Compt. Rend.* **198,** 1154–1156 (1934).

MONDAIN-MONVAL, P., PÂRIS, R. *Compt. Rend.* **207,** 338–339 (1938a).

MONDAIN-MONVAL, P., PÂRIS, R. *Bull. Soc. Chim. France* **5,** Ser. 5, 1641–1655 (1938b).

MONN, D. E. *Diss. Abstr.* **26,** 6998 (1966).

MÜLLER, D. C. Abstr. of Papers, 132nd Meeting, *Am. Chem. Soc.,* Sept., p. 6B (1957).

MÜLLER, R. H. *Ind. Eng. Chem., Anal. Ed.* **13,** 670–671 (1941).

MÜLLER, R. H., STOLTEN, H. *J. Anal. Chem.* **25,** 1103–1106 (1953).

MYLIUS, F. VON, *Z. Metallk.* **14,** 233–244 (1922).

NAKANISHI, M., FUJIEDA, S. *Anal. Chem.* **44,** 574–579 (1972).

NANCOLLAS, G. H., HARDY, J. A. *J. Sci. Instr.* **45,** 290–292 (1967).

NANCOLLAS, G. H., POULTON, D. C. *Inorganic Chem.* **8,** 680–682 (1969).

NAYAR, M. R., PANDE, C. S. *J. Indian Chem. Soc.* **28,** 112–118 (1951).

NELANDER, B. *Acta Chem. Scand.* **20,** 2289–2298 (1966).

NORDON, P., BAINBRIDGE, N. W. *J. Sci. Instr.* **39,** 399 (1962).

NOVOSELOVA, A. V., PASHINKEN, A. S., SEMENKO, K. N. *Vest. Mosk. Univ.* **10,** No. 3; *Fiz. Mat. i Estestven Nauk* **49,** No. 2 (1955).

ORZHEROVSKII, M. A. *Chem. Tech. Oils Fuels* **8,** 61–63 (1969).

PADHYE, V. P. *Analyst* **82,** 634–638 (1957).

PANDYA, P. K., HALDAR, B. C. *J. Sci. Ind. Res., India, B* **21,** 503–504 (1962)

PAPEE, H. M., CANADY, W. J., LAIDLER, K. J. *Can. J. Chem.* **34,** 1677 (1956).

PAPOFF, P., ZAMBONIN, P. G. *Talanta* **14,** 581–590 (1967); PAPOFF, P., TORSI, G., ZAMBONIN, P. G. *Gazz. Chim. Ital.* **95,** 1031–1042 (1965).

PÂRIS, R. A., LE CHATELIER, H. *Compt. Rend.* **199,** 863–865 (1934).

PÂRIS, R. A., ROBERT, J. *Bull. Soc. Chim., France* **10,** 224 (1943).

PÂRIS, R. A., ROBERT, J. *Compt. Rend.* **223,** 1135–1136 (1946).

PÂRIS, R. A., TARDY, P. *Bull. Soc. Chim., France* **12,** 16 (1945).

PÂRIS, R. A., TARDY, P. *Compt. Rend.* **223,** 1001–1003 (1946).

PÂRIS, R. A., VIAL, J. *Chim. Anal., Paris* **34,** 3–5 (1952a).

PÂRIS, R. A., VIAL, J. *Chim. Anal., Paris* **34,** 223–225 (1952b).

PARKER, R. D., VLISMAS, T. *Analyst* **93,** 330–335 (1968).

PARSONS, J. S. Abstr. of Papers, 132nd Meeting, *Am. Chem. Soc.* N.Y., Sept. p. 12B (1957).

PECHAR, F. *Chem. Listy* **59,** 1073–1075 (1965).

PERCHEC, H., GILOT, B. *Bull. Soc. Chim., France,* 619–620 (1964).

PINKSTON, J. L. *Anal. Chem.* **27,** 446–447 (1955).

PITTS, E., PRIESTLEY, P. T. *J. Sci. Instr.* **39,** 75–77 (1962).

POPPER, E., ROMAN, L., MARCU, P. *Talanta* **11,** 515–521 (1964).

POPPER, E., ROMAN, L., MARCU, P. *Talanta* **12,** 249–256 (1965).

POPPER, E., ROMAN, L., MARCU, P. *Talanta* **14,** 1163–1168 (1967).

POULSEN, I., BJERRUM, J. *Acta Chem. Scand.* **9,** 1407–1420 (1955).

PRIESTLEY, P. T. *Analyst* **88,** 194–203; PRIESTLEY, P. T. (1963a); SMITH, M. *Proc. Soc. Anal. Chem.* **3,** 18–19 (1966).

PRIESTLEY, P. T. *J. Sci. Instr.* **40,** 505 (1963b).

PRIESTLEY, P. T. *J. Sci. Instr.* **42,** 35–36 (1965).

PRIESTLEY, P. T., SEBBORN, W. S., SELMAN, R. F. W. *Analyst* **88,** 797–804 (1963); SEBBORN, W. S. *Proc. Soc. Anal. Chem.* **3,** 17–18 (1966).

PRIESTLEY, P. T., SEBBORN, W. S., SELMAN, R. F. W. *Analyst* **90,** 589–593 (1965).

PRIESTLEY, P. T., SEBBORN, W. S., SELMAN, R. F. W. British Patent, 1 109 422 (1968).

PURKAYASTHA, B. C. *J. Indian Chem. Soc.* **24,** 257–262 (1947).

PURKAYASTHA, B. C., SEN-SARMA, R. N. *J. Indian Chem. Soc.* **23,** 31–40 (1946).

QUILTY, C. J. *Anal. Chem.* **39,** 666–668 (1967).

RADMACHER, V. W., SCHMITZ, W. *Brennst. Chemie* **40,** 161–164 (1959).

RÁDY, GY., SAJÓ, I., KAISER, E. *Periodica Polytech.* **11**, 103–110 (1967a).

RÁDY, GY., KAISER, E., GIMESI, O. *Periodica Polytech.* **11**, 111–115 (1967b).

RAFFA, R. J., STERN, M. J., MALSPEIS, L. *Anal. Chem.* **40**, 70–77 (1968).

RAGLAND, J. L. *Soil Sci. Am. Proc.* **26**, 133–137 (1962).

RASMUSSEN, J. L., NIELSEN, T. *Acta Chem. Scand.* **17**, 1623–1629 (1963).

REYNOLDS, C. A., HARRIS, SISTER M. J. *Anal. Chem.* **41**, 348–349 (1969).

RICHARD, H. D., EGGLESTON, J. A. *Analyst* **51**, 281–283 (1926).

RICHARDS, G. W., WOOLF, A. A. *J. Chem. Soc. A*, 1118–1121 (1967).

RICHARDS, G. W., WOOLF, A. A. *J. Chem. Soc. A*, 470–476 (1968).

RICHARDS, G. W., WOOLF, A. A. *J. Chem. Soc. A*, 1072–1076 (1969).

RICHMOND, H. D., MERRYWEATHER, J. E. *Analyst* **42**, 273–274 (1917).

RIVENQ, F. *Bull. Soc. Chim., France* **12**, 283–289 (1945a).

RIVENQ, F. *Bull. Soc. Chim., France* **12**, 289–291 (1945b).

ROMM, I. P., GUR'YANOVA, E. N., GOL'DSHTEIN, I. P., KOCHESHKOV, K. A. *Dokl. Akad. Nauk SSSR* **172**, 618–621 (1967).

ROMM, I. P., GUR'YANOVA, E. N., GOL'DSHTEIN, I. P. *Zh. Fiz. Khim.* **42**, 220–222 (1968).

RONDEAU, J., LEGRAND, M., PÂRIS, R. *Compt. Rend.* **263C**, 579–580 (1966).

ROUGHTON, F. J. W. 'Techniques of Organic Chemistry', Vol. VIII, Part 2, Interscience, New York, p. 758–778 (1963).

RUBY, W. R. *Rev. Sci. Instr.* **26**, 460–462 (1955).

SAJÓ, I. *Kohász. Lapok.* **7**, 287–289 (1957).

SAJÓ, I. *Proc. Anal. Chem. Conf., Budapest*, Vol. III, 445–454 (1966a).

SAJÓ, I. *Kem. Kozlem.* **26**, 119–136 (1966b).

SAJÓ, I. *Talanta* **15**, 578–582 (1968a).

SAJÓ, I. *Z. Anal. Chem.* **242**, 165–172 (1968b).

SAJÓ, I. *Hung. Sci. Instr.* No. 11, 1–8 (1968c).

SAJÓ, I. *Magy. Kem. Folyoirat* **75**, 1–3 (1969a).

SAJÓ, I. *Interceram* **18**, 30–32 (1969b).

SAJÓ, I. *Épitöanyag* **21**, 249–255 (1969c).

SAJÓ, I. *Banyasz. Lapok. Kohaszat* **102**, 258–263 (1969d).

SAJÓ, I., SIPOS, B. *Z. Anal. Chem.* **222**, 23–50 (1966).

SAJÓ, I., SIPOS, B. *Talanta* **14**, 203–213 (1967a).

SAJÓ, I., SIPOS, B. *Mikrochim. Acta*, 248–268 (1967b).

SAJÓ, I., SIPOS, B. *Zement-Kalk-Gips.* **21**, 32–37 (1968a).

SAJÓ, I., SIPOS, B. *Tonind. Ztg. Keram. Rundschau* **92**, 88–92 (1968b).

SAJÓ, I., SIPOS, B. *Kohasz. Lapok* **101**, 484–488 (1968c).

SAJÓ, I., SIPOS, B. *Radex Rundsch.* 178–187 (1968d).

SAJÓ, I., SIPOS, B. Hungarian Patent 155 979 (1969a).

SAJÓ, I., SIPOS, B. *Bull. Soc. Fr. Ceram.* **85**, 3–22 (1969b).

SAJÓ, I., UJVÁRI, J. *Z. Anal. Chem.* **202**, 177–186 (1964).

SAJÓ, I., UJVÁRI, J., SZAMOSKOZI, Z. Hungarian Patent 152 942 (1966).

SAJÓ, I., FOESTER, W., REUDIGER, H., SIPOS, B. *Neue Hütte* **12**, 500–502 (1967).

SAJÓ, I., SIPOS, B., KLUG, O. *Kohasz. Lapok.* **102**, 89–91 (1969).

SAXENA, R. S., BHATTACHARYA, A. K. *J. Indian Chem. Soc.* **28**, 703–709 (1951).

SAXENA, R. S., BHATTACHARYA, A. K. *J. Indian Chem. Soc.* **29**, 632–635 (1952).

SCHÄFER, H., WILDE, E. *Z. Anal. Chem.* **130**, 396–402 (1949).

SCHLYTER, K. *Trans. Roy. Inst. Technol., Stockholm* **132**, 40 pp (1959).

SCHLYTER, K., SILLÉN, L. G. *Acta Chem. Scand.* **13**, 385–386 (1959).

SEN, B., WU, W. C. *Anal. Chim. Acta* **46**, 37–47 (1969).

SIDDHANTA, S. K. *Sci. Cult., Calcutta* **12**, 409–410 (1947).

SIDDHANTA, S. K. *J. Indian Chem. Soc.* **25**, 579–583 (1948a).

SIDDHANTA, S. K. *J. Indian Chem. Soc.* **25**, 585–588 (1948b).

SIDDHANTA, S. K., GUHA, M. P. *J. Indian Chem. Soc.* **32**, 355–365 (1955).

SIJDERIUS, R. *Anal. Chim. Acta* **10**, 517–519 (1954).

SILLÉN, L. G., MARTELL, A. E. 'Stability Constants', The Chemical Society, London, Special Publication No. 17 (1964).

SIPOS, V., SAJÓ, I. *Proc. Anal. Chem. Conf., Budapest*, Vol. III, 455–461 (1966).

SMITH, J. B. *Diss. Abstr. Int. B* **30**, 92 (1969).

SMITH, T. B. 'Analytical Processes', E. Arnold, London (1952).

SNELSON, F. L., ELLIS, W. R., VILKAULS, J. *Analyst* **92**, 264–267 (1967).

SNYDER, K. L. *Chem. Eng. Progr.* **64**, 75–78 (1968).

SOMIYA, T. *J. Soc. Chem. Ind. (Japan)* **30**, 106–112 (1927a).

SOMIYA, T. *Proc. Imp. Acad. (Japan)* **3**, 76–78 (1927b).

SOMIYA, T. *J. Soc. Chem. Ind. (Japan)* **31**, 306–311 (1928a).

SOMIYA, T. *Chem. News* **137**, 14 (1928b).

SOMIYA, T. *Proc. Imp. Acad. (Japan)* **5**, 34–37 (1929a).

SOMIYA, T. *J. Soc. Chem. Ind. (Japan)* **32**, 490–495 (1929b).

SOMIYA, T. *J. Soc. Chem. Ind. (Japan)* **32**, *Suppl. binding*, 153–154 (1929c).

SOMIYA, T. *J. Soc. Chem. Ind. (Japan)* **33**, 140–142 (1930a).

SOMIYA, T. *J. Soc. Chem. Ind. (Japan)* **33**, *Suppl. binding*, 174–176 (1930b).

SOMIYA, T. *J. Soc. Chem. Ind. (Japan)* **34**, *Suppl. binding* 281–282 (1931).

SOMIYA, T. *J. Soc. Chem. Ind.* **51**, 135–140 (1932).

SPINK, M. Y., SPINK, C. H. *Anal. Chem.* **40**, 617–619 (1968).

SPITSYN, V. I., KOSMODEM'YANSKAYA, G. V. *Zh. Neorgan. Khim.* **10**, 353 (1965).

SPITSYN, V. I., KOSMODEM'YANSKAYA, G. V. *Omagiu Raluca Ripan*, 581–588 (1966a).

SPITSYN, V. I., KOSMODEM'YANSKAYA, G. V. *Zh. Neorgan. Khim.*, **11**, 1397–1400 (1966b).

SPITSYN, V. I., KOSMODEM'YANSKAYA, G. V. *Zh. Neorgan. Khim.* **13**, 2213–2215 (1968).

STERN, M. J., WITHNELL, R., RAFFA, R. J. *Anal. Chem.* **38**, 1275–1277 (1966).

S.T.P.T.C. 'Standard methods for testing tar and its products', 6th Edition, S.T.P.T.C., Gomersal, Cleckheaton, Yorkshire (1967).

ŠTRÁFELDA, F. Czechoslov Patent 127 963 (1968).

ŠTRÁFELDA, F., HÁJKOVÁ, M. *Collection Czech Chem. Commun.* **33**, 4389–4392 (1968).

ŠTRÁFELDA, F., KROFTOVÁ, J. *Collection Czech. Chem. Commun.* **33**, 3694–3702 (1968a).

ŠTRÁFELDA, F., KROFTOVÁ, J. *Collection Czech. Chem. Commun.* **33**, 4171–4177 (1968a).

SUNNER, S., WADSÖ, I. *Acta Chem. Scand.* **13**, 97–108 (1959).

SYKES, P. 'A Guidebook to Mechanisms in Organic Chemistry', 5th Edition, Longmans, Green & Co. Ltd., London, p. 158 (1963).

TAKENO, R. *Ryusan* **17**, 253–263 (1964).

TAKENO, R. *Ryusan* **18**, 211–217; 231–235 (1965).

TAKEUCHI, T., YAMAZAKI, M. *Kogyo Kagaku Zasshi* **72**, 1263–1269 (1969a).

TAKEUCHI, T., YAMAZAKI, M. *Kogyo Kagaku Zasshi* **72**, 1500–1503 (1969b).

TAUBINGER, R. P. *Analyst* **94**, 634–642 (1969a).

TAUBINGER, R. P. *Soc. Anal. Chem.*, Automatic Methods Group, Demonstration meeting, April (unpublished work) (1969b).

TAYLOR, M. P. *Analyst* **80**, 153–155 (1955).

TEJAM, B. M., HALDAR, B. C., *Indian J. Chem.* **7**, 613–615 (1969).

THIESSE, X. *Bull. Soc. Chim. France* **7**, 495 (1940).

TRAMBOUZE, Y., PASCAL, P. *Compt. Rend.* **233**, 648–649 (1951).

TRAMBOUZE, Y., MOURGUES, L. DE, PERRIN, M. *Compt. Rend.* **234**, 1770–1772 (1952).

TYRRELL, H. J. V. *Talanta* **14**, 843–848 (1967).

TYRRELL, H. J. V., BEEZER, A. E. 'Thermometric Titrimetry', Chapman and Hall Ltd., London (1968).

TYSON, B. C., Jr., MCCURDY, W. H., Jr., BRICKER, C. E. *Anal. Chem.* **33**, 1640–1645 (1961).

VAJGAND, V. J., GAÁL, F. F. *Glasnik. Hem. Drustva, Beograd*, **31**, 103–120 (1966).

VAJGAND, V. J., GAÁL, F. F. *Talanta* **14**, 345–351 (1967).

VAJGAND, V. J., PASTOR, T., TODOROVSKI, T., GAÁL, F. F., TODOROVIĆ, M. *Proc. Anal. Chem. Conf., Budapest*, 152–158 (1966).

VAJGAND, V. J., KISS, T. A., GAÁL, F. F., ZSIGRAI, I. J. *Talanta* **15**, 699–704 (1968).

VAJGAND, V. J., GAÁL, F. F., BRUSIN, S. S. *Talanta* **17**, 415–421 (1970).

VAJGAND, V. J., GAÁL, F. F., ZRNIÓ, LJ., BRUSIN, S. S., VELIMIROVIC, D. To be published (1971).

VALCHA, J. *Chem. Prumysl* **15**, 397–401 (1965).

VAN DER HEIJDE, H. B., DAHMEN, E. A. M. F. *Anal. Chim. Acta* **16**, 378–400 (1957).

VANDERSEE, C. E., SWANSON, J. A. *J. Phys. Chem.* **67**, 2609 (1963).

VARTAK, D. G., KABADI, M. B. *J. Univ. Bombay* **22**, Pt. 5, Sci. No., Sect. A, No. 35, 34–39 (1954).

VARTAK, D. G., KABADI, M. B. *J. Indian Chem. Soc.* **32**, 351–354 (1955).

VAUGHAN, G. A. Report No. 0364, Coal Tar Research Association, Gomersal (1966).

VAUGHAN, G. A. British Wood Preserving Association, Cambridge Meeting, 16–28 (1967).

VAUGHAN, G. A. Unpublished work (1970).

VAUGHAN, G. A., SWITHENBANK, J. J. *Analyst* **90,** 594–599 (1965); VAUGHAN, G. A. *Proc. Soc. Anal. Chem.* **3,** 20 (1966).

VAUGHAN, G. A., SWITHENBANK, J. J. *Analyst* **92,** 364–370 (1967).

VAUGHAN, G. A., SWITHENBANK, J. J. *Analyst* **95,** 890–893 (1970).

VOGEL, A. I. 'A Textbook of Quantitative Inorganic Analysis', Longmans, London (1964).

WADSÖ, I., MONK, P. *Acta Chem. Scand.* **22,** 1842–1852 (1968).

WANDERS, A. C. M., ZWEITERING, Th. N. *J. Phys. Chem.* **73,** 2076–2078 (1969).

WASILEWSKI, J. C. U.S. Patent 3 106 477 (1964).

WASILEWSKI, J. C., MILLER, C. D. *Anal. Chem.* **38,** 1750–1751 (1966).

WASILEWSKI, J. C., PEI, P. T., JORDAN, J. *Anal. Chem.* **36,** 2131–2133; Abstr. of Papers, 147th Meeting *Am. Chem. Soc.*, Philadelphia, p. 12B (1964).

WATT, G. D. *Diss. Abstr.* **27B,** 1406 (1966).

WEINER, N. D., FELMEISTER, A., *Anal. Chem.* **38,** 515–516 (1966).

WEISZ, H., KISS, T. A. *Z. Anal. Chem.* **249,** 302–303 (1970).

WEISZ, H., KISS, T. A., KLOCKOW, D. *Z. Anal. Chem.* **247,** 248–252 (1969).

WILLARD, H. H., MERRITT, L. L., DEAN, J. A. 'Instrumental Methods of Analysis', 4th Edition, Van Nostrand, Princeton, N.J. pp. 470–472 (1965).

WILLIAMS, M. B., JANATA, J. *Talanta* **17,** 548–551 (1970).

WILLS, V., STAMBURG, O. F., PROCTOR, Z. G. *Anal. Chem.* **34,** 224–225 (1962).

WILSON, E. W., Jr., SMITH, D. F. *Anal. Chem.* **41,** 1903 (1969).

WRATHALL, D. P., IZATT, R. M., CHRISTENSEN, J. J. *J. Am. Chem. Soc.*, **86,** 4779 (1964).

YAGUB'YAN, E. S., BUKHALOVA, G. A., KHILYAN, T. M. *Zh. Neorgan. Khim.* **10,** 2581–2583 (1965).

ZAMBONIN, P. G., JORDAN, J. *Anal. Chem.* **41,** 437–442 (1969).

ZENCHELSKY, S. T., SEGATTO, P. R. Abstr. of Papers, 132nd Meeting, *Am. Chem. Soc.* p. 12B (1957a).

ZENCHELSKY, S. T., SEGATTO, P. R. *Anal. Chem.* **29,** 1856–1858 (1957b).

ZENCHELSKY, S. T., SEGATTO, P. R. *J. Am. Chem. Soc.* **80,** 4796–4799 (1958).

ZENCHELSKY, S. T., PERIALE, J., COBB, J. C. *Anal. Chem.* **28,** 67–69 (1956).

Index